AF303156

aus der Reihe:

Innovationen mit Mikrowellen und Licht

**Forschungsberichte aus dem Ferdinand-Braun-Institut
Leibniz-Institut für Höchstfrequenztechnik**

Band 23

Ponky Ivo

AlGaN/GaN HEMTs Reliability:
Degradation Modes and Analysis

Herausgeber: Prof. Dr. Günther Tränkle, Prof. Dr.-Ing. Wolfgang Heinrich

Ferdinand-Braun-Institut
Leibniz-Institut
für Höchstfrequenztechnik (FBH)
Gustav-Kirchhoff-Straße 4
12489 Berlin

Tel.   +49.30.6392-2600
Fax   +49.30.6392-2602

E-Mail fbh@fbh-berlin.de
Web   www.fbh-berlin.de

Innovations with Microwaves and Light

Research Reports from the Ferdinand-Braun-Institut, Leibniz-Institut für Höchstfrequenztechnik

Preface of the Editors

Research-based ideas, developments and concepts are the basis of scientific progress and competitiveness, expanding human knowledge and being expressed technologically as inventions. The resulting innovative products and services eventually find their way into public life.

Accordingly, the *"Research Reports from the Ferdinand-Braun-Institut, Leibniz-Institut für Höchstfrequenztechnik"* series compile the institute's latest research and developments. We would like to make our results broadly accessible and to stimulate further discussions, not least to enable as many of our developments as possible to enhance everyday life.

Reliable device performance is the key for successful implementation of GaN devices for microwave and high voltage power applications. A proper understanding of degradation mechanisms provides the prerequisite for a focused engineering of reliability into the devices. Therefore, this thesis concentrates on methods to spot degradation effects in a very initial state (statu nascendi) in order to study responsible physical mechanisms. In this connection a step stressing method combined with electrical measurements and electroluminescence analyses has been established and applied to GaN test transistors fabricated on different epitaxial materials and by using different processing approaches. It could be found that degradation starts in very distinct device areas spreading all over the device when stressing increases. The threshold for degradation depends on specific material design and growth conditions as well as on processing and layout details. In general, the work figures out basic initial degradation effects of GaN devices and gives hints on engineering devices for improved reliability.

We wish you an informative and inspiring read.

Prof. Dr. Günther Tränkle
Director

Prof. Dr.-Ing. Wolfgang Heinrich
Deputy Director

The Ferdinand-Braun-Institut

The Ferdinand-Braun-Institut researches electronic and optical components, modules and systems based on compound semiconductors. These devices are key enablers that address the needs of today's society in fields like communications, energy, health and mobility. Specifically, FBH develops light sources from the visible to the ultra-violet spectral range: high-power diode lasers with excellent beam quality, UV light sources and hybrid laser systems. Applications range from medical technology, high-precision metrology and sensors to optical communications in space. In the field of microwaves, FBH develops high-efficiency multi-functional power amplifiers and millimeter wave frontends targeting energy-efficient mobile communications as well as car safety systems. In addition, compact atmospheric microwave plasma sources that operate with economic low-voltage drivers are fabricated for use in a variety of applications, such as the treatment of skin diseases.

The FBH is a competence center for III-V compound semiconductors and has a strong international reputation. FBH competence covers the full range of capabilities, from design to fabrication to device characterization.

In close cooperation with industry, its research results lead to cutting-edge products. The institute also successfully turns innovative product ideas into spin-off companies. Thus, working in strategic partnerships with industry, FBH assures Germany's technological excellence in microwave and optoelectronic research.

AlGaN/GaN HEMTs Reliability: Degradation Modes and Analysis

vorgelegt von
Master of Physics
Ponky Ivo
aus Jakarta

von der Fakultät IV - Elektrotechnik und Informatik
der Technische Universität Berlin
zur Erlangung des akademischen Grades

Doktor der Naturwissenschaften
-Dr.rer.nat.-

genehmigte Dissertation

Promotionsausschuss:

Vorsitzender: Prof. Dr.-Ing. W. Heinrich
Berichter: Prof. Dr. G. Tränkle
Berichter: Prof. Dr.-Ing. C. Boit
Berichter: Prof. Dr. M. Kuball

Tag der wissenschaftliche Aussprache: 24. Mai 2012

Berlin 2012
D 83

Bibliografische Information der Deutschen Nationalbibliothek
Die Deutsche Nationalbibliothek verzeichnet diese Publikation in der Deutschen Nationalbibliografie; detaillierte bibliografische Daten sind im Internet über http://dnb.d-nb.de abrufbar.
1. Aufl. - Göttingen: Cuvillier, 2012
Zugl.: (TU) Berlin, Univ., Diss., 2012

978-3-95404-259-3

© CUVILLIER VERLAG, Göttingen 2012
Nonnenstieg 8, 37075 Göttingen
Telefon: 0551-54724-0
Telefax: 0551-54724-21
www.cuvillier.de

Acknowledgments

I would like to thank Professor Dr. Günther Tränkle for giving me the chance to work at Ferdinand-Braun-Institut für Höchstfrequenztechnik (FBH). His guidance through discussions not only encourage me to work and dig more GaN reliability studies, but also build up my knowlegde and broaden my horizon.

A respectful thank I would like to address to Dr. Joachim Würfl for fruitful discussions and his willingness to share his broad knowledge and long experience in GaN areas. I highly apreciate his hardworking attitude and nurture skills which are very good examples for me professionally and personally.

I am very grateful to Dr. Richard Lossy to whom I can ask a lot of questions regarding GaN process technology and reliability problems.

My learning process in Germany would not be possible without DAAD funding. I am grateful to Fr. Kasperek and Mbak Endah for their assistance and guidance prior to my departure from Jakarta as well as my stay in Berlin.

During my PhD work, there were some collaborations on GaN reliability studies. Therefore, I would to express my gratitude to Professor Christian Boit group at Technische Universität (TU) Berlin, Professor Martin Kuball and Dr. James Pomoroy at University of Bristol, Professor Gaudio Meneghesso at University of Padova, and Dr. Tim and Fr. Sandy Schaaf at Max Born Institute.

Special thanks to Dr. Arkadiuzs Glowacki at TU Berlin for teaching me how to use PEM machine. Thank you very much for your trust and sharing your lab skills. I would like to thank to Dr. Ute Zeimer for her assistance in analyzing the samples and discussing the results of EDX, CL and FIB. Thanks also to Fr. Lawrenz for FIB-ing assistance. And to Dr. Anna Mogilatenko and Dr. Holm Kirmse at Humboldt University, I thank for their corporation and fruitful discussions concerning FIB, lamella transfer, EDX and TEM investigations. Special thanks to Anna for her valuable corrections for my thesis.

Thanks to ex-*Büro* colleagues: Reza, Paul, Tomas and Mathias. Special thanks to Nidhi for her useful tips for academic matters and for living in Germany happilly. Eldad Bahat-Treidel, thank you for wafers supply and simulations (you surprised me with Indonesian greeting in the first day I worked!). And Kotara, many thanks for your efforts in FIB and discussions. Thanks for inputs and discussions to other GaN group members. To Melanie, thanks for your willingness to share your feedbacks, findings and feelings as well.

My four year work cannot be accomplished without help from wonderful measurement guys: Lars who is always cooperative and very helpful, Stephan and Steffen who nearly don't mind their DIVA measurements to be interrupted, Hai Bang and Marko who almost do *Aufbau* work quickly for me. Also thanks to Armin, for your assitance and sharing your long experience. It is also good to work with nice persons in bending measurements so I thank Hr. Roos, Ben and Irene.

I would like also to address my sincere gratitude to Claudia, *du bist ein Schutzengel*. Thanks to Ngoc for her companion working in the weekend. I thank other Ladies Night (LN) members of FBH. Thanks to Sabir, Aga, Olof and Thomas for free advices which most of them make me laugh :D.

I am indebted to friends who are always there for me for good and bad times in DE especially: Ican, Yuni and Rere, Jeung Lyza and Mas Ken, Eva, Ratri, Riska. Special thanks to friends who always support me no matter how far we are: Rientha, Lina, Yayah, Rosie, Mutiara and Yenni for their patience in listening to my stories. To Alex and Bernt, thanks for your understanding during my stressful time. Thanks also to my friends in Groningen: Uyung, Puri, Amel and Iging for sharing our *desperado* period during PhD and supporting each other. Thanks to *Mesjid* Al-Falah (IWKZ) where I can satisfy needs of my soul and stomach. Thanks to DIBVM where I enjoy playing badminton every saturday and to DAAD 2007 members for sharing our experience and support. Special thanks to Bang Aji for his important contribution which brightens my life in Berlin.

My study far away from home would not be blessed without my parents' prayers. I thank Allah, The Almighty, for giving me wonderful parents and a warm big family. Thank you very much for your endless support, love and encouragements.

Berlin, February 2012

Abstract

AlGaN/GaN HEMTs reliability and stability issues were investigated in dependence on epitaxial design and process modification. DC-Step-Stress-Tests have been performed on wafers as a fast device robustness screening method. As a criterion of robustness they deliver a critical source-drain voltage for the onset of degradation. Several degradation modes were observed which depend on epi design, epi quality and process technology. Electrical and optical characterizations together with electric field simulations were performed to get insight into respective degradation modes. It has been found that AlGaN/GaN HEMT devices with GaN cap show higher critical source-drain voltages as compared to non-capped devices. Devices with low Al concentration in the AlGaN barrier layer also show higher critical source-drain voltages. Superior stability and robustness performance have been achieved from devices with AlGaN backbarrier epi design grown on n-type SiC substrate. For the onset on any degradation modes the presence of high electrical fields is most decisive for on- and off-state operation conditions. Therefore careful epi design to reduce high electric field is mandatory. It is also shown that epi buffer quality and growth process have a great impact on device robustness. Defects such as point defects and dislocations are assumed to be created initially during stressing and accumulated to larger defect clusters during device stressing. Electroluminescence (EL) measurements were performed to detect early degradation. Extended localized defects are resulting as bright spots at OFF-state conditions in conjunction with a gate leakage increase.

Zusammenfassung

AlGaN/GaN HEMTs mit unterschiedlichen epitaktischen Designs und Prozess-modifikationen wurden auf ihre Zuverlässigkeit und Stabilität untersucht. DC-Stufenstresstests wurden als Screeningmethode für die Bauelementrobustheit durchgeführt. Mit dieser Methode erhält man eine kritische Source-Drain-Spannung, die den Beginn der Degradation kennzeichnet. Verschiedene Degradationsmodi wurden beobachtet, die vom epitaxialem Design, der epitaxialen Qualität und der Prozesstechnologie abhängen. Elektrische und optische Messungen zusammen mit elektrischen Feldsimulationen wurden durchgeführt, um Einblick in das Degradationsverhalten zu bekommen. Es hat sich gezeigt, dass AlGaN/GaN HEMTs mit einer GaN Cap-Schicht eine höhere kritische Drain-Source-Spannung zeigen als Transistoren ohne diese Schicht. HEMTs mit niedriger Aluminiumkonzentration in der AlGaN-Barriere zeigen ebenfalls eine höhere kritische Drain-Source-Spannung. Transistoren mit AlGaN-Backbarrier, die auf n-Typ SiC-Substraten gewachsen wurden, zeigen eine besonders hohe Stabilität und Robustheit. Für den Betrieb im On-State als auch im Off-State ist ein hohes elektrisches Feld entscheidend für den Beginn der Degradation. Daher sind epitaxiale Designs, die das elektrische Feld so weit wie möglich reduzieren, von großer Wichtigkeit. Es wird gezeigt, dass die Qualität der Bufferschicht und der Wachstumsprozess der epitaxierten Schichten großen Einfluß auf die Robustheit der Bauelemente haben. Zu Beginn des Stressprozesses werden Punktdefekte und Versetzungen erzeugt, die im weiteren Verlauf des Stresstests zu Agglomeration von Defektclustern führen. Der Beginn der Degradation wurde mit Hilfe der Elektrolumineszenz untersucht. Im Off-State werden ausgedehnte lokalisierte Defekte als stark leuchtende Flecken detektiert, wobei gleichzeitig ein Anstieg der Leckströme zu beobachten ist.

Author's Declaration

This thesis is submitted to Technische Universität Berlin in support of an application for admission to the degree of Dr. rer. nat. I hereby declare that this thesis is my own work and effort and that it has not been submitted anywhere for any award. Where other sources of information have been used, they have been acknowledged. The work was carried out between September 2007 and February 2012, under supervision of Professor Günther Tränkle.

Ponky Ivo
February 2012

Contents

Chapter 1

Introduction

1.1   A short story of GaN

The first gallium nitride (GaN) compounds were prepared by reacting ammonia gas with metallic gallium at high temperatures in 1932 by Johnsonn *et al.* [1]. Their work showed its remarkable stability toward heat, solutions of acids and bases. Almost four decades later Maruska *et al.* [2] and Pankove *et al.* [3] characterized the optical properties of GaN and determined the direct band gap to 3.36 - 3.39 eV at room temperature. A few years later the GaN bandstructure and its reflectivity were computed by the empirical pseudopotential method [4]. Bloom *et al.* predicted GaN as a promising material for luminescence devices and laser applications due to its wide direct band energy. In the following years, Ilegems and Montgomery [5] suggested that the n-type conductivity of GaN semiconductor behaviour is due to native defects. Their conclusions were widely debated. The question was whether n-type conductivity was due to native defects or impurities. Several groups continued their work intensively on point defects and most comprehensive point defects studies are contributed by Neugebauer and Van de Waal [6]. They proposed that unintentional donor impurities and gallium vacancies are responsible for n-type conductivity of GaN rather than the long believed-nitrogen vacancy [7]. Other noticable work on charged point defects which control numerous defect properties of semiconductors have been published by Seebauer and Kratzer [8].

In early 90's, the first p-n junction LED GaN fabrication was reported by a Japanese group [9]. Earlier reports of improving crystalline quality of GaN by inserting a thin AlN nucleation layer to accomodate lattice constant mismatch of sapphire substrate was reported by Yoshida *et al.* and Akasaki *et al.* [10, 11]. In early 90's Khan *et al.* for the first time observed a 2-dimensional electron gas (2DEG) at the interface between AlGaN and GaN layers [12]. Since then, research of AlGaN/GaN high electron mobility transistors (HEMTs) towards improving technology for improved performance and novel application was intensively pursued by numbers of groups [13, 14, 15, 16, 17]- to mention a few.

Mishra's recent overview provided impressive data of RF performance of GaN HEMTs stand: 13.7 W/mm at 30 GHz, 10.5 W/mm at 40 GHz, and 2.4 W/mm at 60 GHz and the fastest GaN devices today at a cut-off frequency of 220 GHz and a maximum oscillation frequency of 400 GHz [18]. However, GaN-based devices have shown short-term instabilities including collapse of DC IV-characteristics, and high leakage currents. This urges comprehensive reliability studies which are the main purpose of this thesis. It is important to investigate the main cause of GaN HEMTs degradation i.e. temperature and/or electric field dependencies through detailed investigations from material growth to process technology.

1.2   GaN structure

Group III-nitrides can crystallize in three possible crystal structures: the wurzite (WZ) structure, the zinc-blende structure and the the rocksalt structure. The GaN WZ structure has a hexagonal unit cell and is thermodynamically more stable than other structures such as zinc blende or rocksalt structures [19]. The wurtzite (WZ) structure is non-centrosymetric (i.e. it lacks of inversion symmetry) and displays piezoelectric effect. The asymmetry of Ga-N bonding between the longer bond and the shortest bond in a tetrahedral atom arrangement leads to a permanent dipole along the c-axis (see Fig. 1.1b). Ga-N bond is highly polarized with the electrons located mostly near the nitrogen atom [20] (see Fig. 1.1c). The spontaneous polarization of GaN crystal by convention is along $[000\bar{1}]$ direction [21] (see Fig. 1.2). Any stress accomodated during the heteroepitaxial GaN growth changes the lattice parameter along the c-axis, leading to an additional piezoelectric polarization. Piezoelectric constant of AlN, GaN and InN with WZ structure have are up to ten times larger than those of conventional III-V and II-VI semiconductor compound [22].

1.3   Substrate options

The sucessful growth of GaN epitaxial layer on foreign substrates has to consider several aspects: lattice mismatch, thermal conductivity and price.

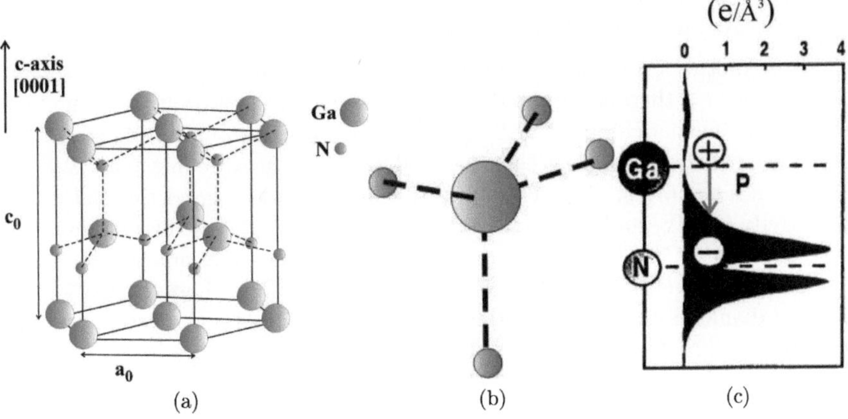

Figure 1.1 GaN (a) wurtzite structure, (b) tetrahedral configuration, and (c) charge distribution of valence electrons.

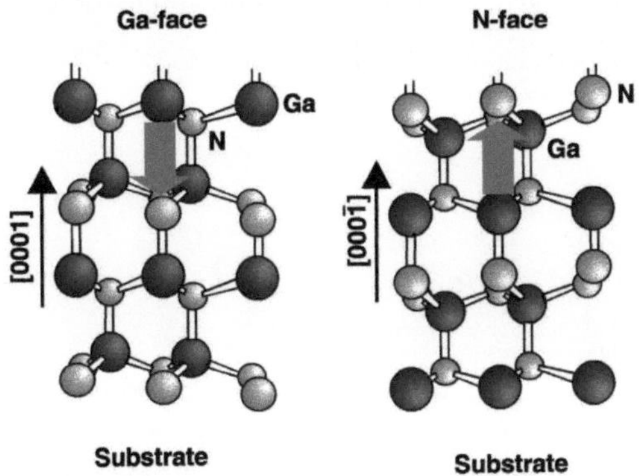

Figure 1.2 Ga-polarity and N-polarity crystals with each polarization direction (blue arrow)[13].

Historically, sapphire (Al_2O_3) was the first substrate for GaN-based devices, but it is not that favorable for RF and high power device applications due to its poor thermal conductivity (0.2-0.5 W/cm K). Moreover, GaN growth on sapphire has big lattice mismatch 13-16 % [23, 24]. The most favorable GaN substrate for microwave applications is SiC which is good in terms of thermal conductivity of 5 W/cm K, less expensive and provides a comparable low lattice mismatch ~ 3.4 % [25]. Growing GaN-based device on SiC substrate with AlN nucleation layer, the strain can be smaller ~ 0.5 % [26].

Another option are Si substrates which are cheaper than SiC substrates and have a thermal conductivity of 1.5 W/cm K. Disadvantage is the higher lattice mismatch about 16.9 % [27] which causes strong tensile strain. The best substrate to grow GaN-based device is a freestanding GaN substrate with advantages, of course, no lattice mismatch and good thermal conductivity (1.3 W/cm K). Recently, it was reported a very low thermal impedance of AlGaN/GaN HEMTs on diamond substrate is 4.1 K mm/W due to high diamond thermal conductivity of 22 W/cm K) [28]. However, GaN growth on diamond has large lattice mismatch which causes severe strain and wafer bowing [29] (see Table 1.1).

Fig. 1.3 shows estimations of the annual wafer production for each substrate type. For large volume production the Si substrate is the most attractive option for power electronics applications. Fre standing GaN substrates are good for optoelectronic applications since they are very sensitive to vertical dislocations. For microelectronic applications, still SiC substrate is favorable.

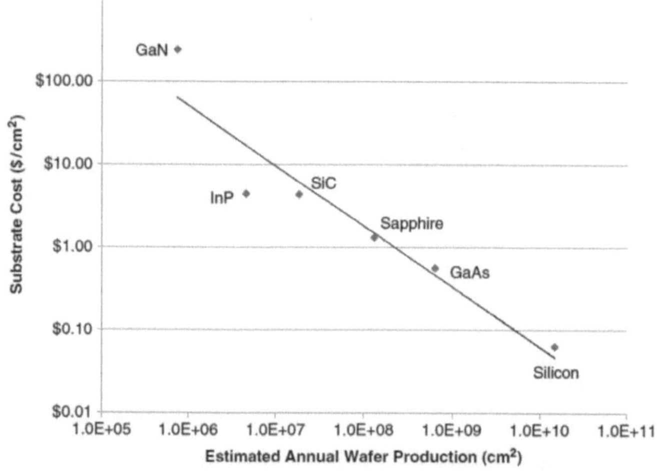

Figure 1.3 Substrate cost per square centimeter estimation in annual wafer production [30].

Table 1.1 Summary of alternative to SiC substrate for GaN-based device [29]

Substrate	Advantage	Disadvantage
GaN	lower relative defect density, lower leakage current, higher device yield, good for optoelectronic applications	difficult to produce large-diameter single-crystal GaN substrate, expensive
Diamond	high thermal conductivity, higher operating power density and temperature capabilities	large lattice mimatch causes severe strain and bowing, not easy to deposit GaN
Si	offers integration with Si IC technologies, large diameter substrate	lower thermal conductivity, big lattice mismatch

1.4 AlGaN/GaN high electron mobility transistors (HEMTs)

AlGaN/GaN HEMTs are fabricated at FBH by metal organic vapour phase epitaxy (MOVPE) in crystal direction (0001) with gallium face surface. On the top of the GaN buffer layer, a thin AlGaN layer is deposited. At the interface of this heterostructure, a 2DEG is formed (see Fig. 1.4). When two different semiconductors are in contact in the absence of an external bias voltage, the equilibrium is reached by lining up the Fermi level and bending the band diagram accordingly. This can create triangully shaped quantum well structure at the interface. The necessary band bending is a consequent to the transfer of electrons from semiconductor with larger band gap (i.e. AlGaN) to lower band gap (i.e. GaN). The transfer of electrons to occupy lower energy states continues until the Fermi level is the same on both sides of heterostructures [31, 32]. The electrons occupy energy states in the triangular potential well the so called 2DEG such that electrons have quantized energy and are free to move in a two dimensional plane parallel to the interface.

The AlGaN crystal has a smaller lattice constant than GaN crystal because the Ga atom in GaN crystal is replaced by smaller sized Al atom. Consequently, the AlGaN a-axis lattice spacing must stretch to match to the underlying GaN lattice (see Fig. 1.5). This causes the c-axis of the AlGaN layer to contract and the lattice is unrelaxed and an additional piezoelectric polarization P_{PE} in AlGaN layer occurs. The total polarization field of both spontaneous polarization P_{SP} and piezoelectric polarization P_{PE} induces charges in 2DEG at the AlGaN/GaN interface with a high sheet charge density of $\sim 10^{13}$ cm^{-2}. It was shown that there is a minimum AlGaN thickness necessarily to induce charges in 2DEG [33] and that the Al concentration in

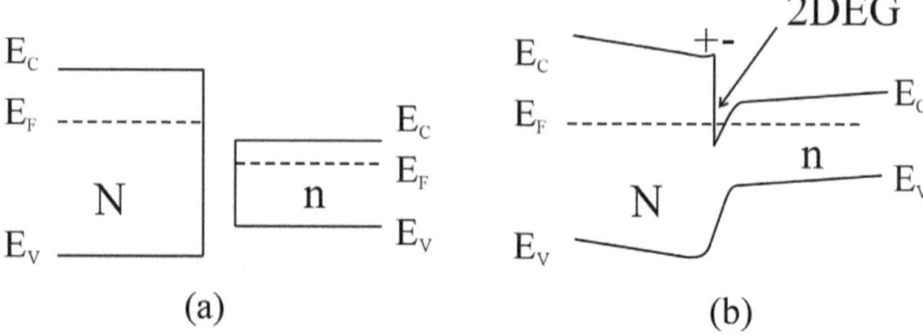

Figure 1.4 Schematic formation of 2DEG in the interface of heterostructure semi-conductors due to the conduction band discontinuity in the interface (a) before contact, and (b) in contact reaching the equilibrium by lining up Fermi level. Electrons transfer from larger band gap semiconductor (N) to lower band gap semiconductor (n) results a positive donor space charge in larger band gap semiconductor.

AlGaN layer determines the charge sheet density in the 2DEG channel [34]. Fig. 1.6 depicts the composition-dependent strain and the total polarization for a thin $Al_xGa_{1x}N$ layer on a relaxed GaN layer.

The charge carriers in 2DEG channel due to total polarization ($P = P_{SP} + P_{PE}$) is described by Poisson equation as the following [29],

$$\nabla \cdot D = \nabla \cdot (\varepsilon\, E + P) = \rho$$
$$\nabla \cdot [\varepsilon(-\nabla\varphi)] + \nabla \cdot P = \rho$$
$$\nabla\, \varepsilon \to 0, \text{ then}$$
$$\nabla^2\varphi = -\frac{\rho}{\varepsilon} + \frac{1}{\varepsilon}\,[\nabla \cdot P],$$

At the interface between AlGaN and GaN, the polarization results in a fixed polarization charge ρ^{Pol}

$$\nabla^2\varphi = -\frac{\rho}{\varepsilon} - \frac{\rho^{Pol}}{\varepsilon}$$

The charge ρ in the semiconductor covers holes p, electrons n, ionized donors N_D^+, and ionized acceptors N_A^-; and possibly the donor-acceptor traps. Thus a complete description of charged carried in semiconductor due to the response of a potential field as the following,

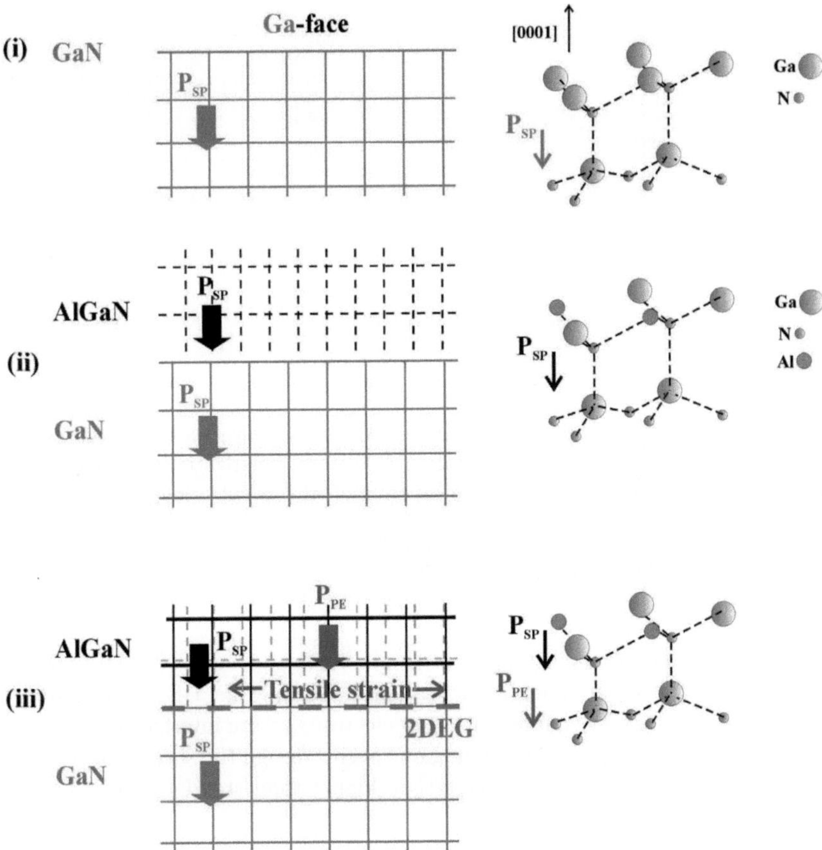

Figure 1.5 Schematic explanation of polarization induced charge creation (i) GaN growth with Ga-face has spontaneous polarization field P_{SP} along c-axis represented by a blue arrow (ii) before intimate contact- AlGaN layer with smaller lattice constant (dashed black lines) than GaN lattice constant due to smaller size of Al atom (iii) deposition of AlGaN layer on the top of GaN layer creates tensile strain along a-axis to match GaN lattice constant and consequently the lattice constant of AlGaN in c-axis is stretched. This creates piezoelectric polarization P_{PE} in AlGaN layer.

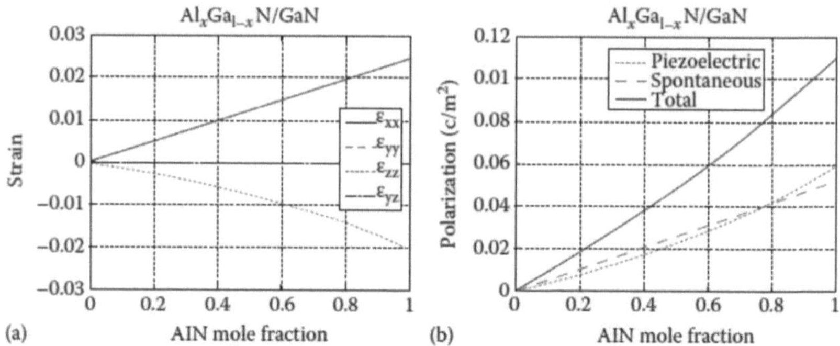

(a)   (b)

Figure 1.6 (a) Elastic strain calculation in Al$_x$Ga$_{1-x}$N layer as a function of Al concentration on a relaxed GaN crystal, and (b) total polarization from piezoelectric and spontaneous polarizations as a function of Al concentration [29].

$$\nabla^2 \varphi = - \frac{q}{\varepsilon} \left[N_D^+ - N_A^- + p - n \right] - \frac{\rho^{Pol}}{\varepsilon}$$

Fig. 1.7 describes high concentration of charged carries in 2DEG as a response to the potential field. The standard AlGaN/GaN HEMTs is characterized by a high sheet charge density of electrons at the interface even without any intentional doping. Depending on specific design the sheet charge density can be almost one order of magnitude higher than standard AlGaAs/GaAs HEMTs.

1.5   The status of GaN reliability

Literature of GaN device reliability has some decisive questions: what is the real cause of GaN degradation? Is it due to material intrinsic properties, substrate and/or GaN growth quality, process related effects or a combination of all of them? The most important question is: what is physics behind it?

Presently, there are two main streams of degradation mechanism explanations: hot electron [35, 36] and inverse piezoelectric effect [37]. Hot electrons by definition are electrons with higher energy than the lattice thermal energy. They can get kinetic energy from high electric field when the device in turned on. These "hot" electrons can get injected into the AlGaN barrier layer and might be trapped or create interface states or bulk traps [35]. Defect creation by hot electrons in GaN and other similar materials is controlled

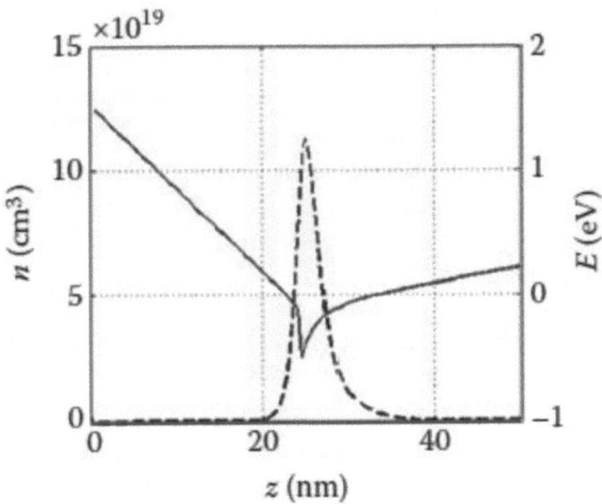

Figure 1.7 Charged carriers (dashed line) concentration in 2DEG channel of standard conduction band (solid line) AlGaN/GaN HEMTs as a response to the potential field [29].

by local kinetic considerations, and not by the defect formation energy. Hot electrons can possibly provide sufficient energy to cause a pre-existing defect to convert into a metastable configuration or cause migration of pre-existing defects. In addition, hot electrons may release a hydrogen atom from a pre-existing passivated defect. Hydrogen release (possibly from annealing) from defects has been known to cause device degradation in Si [38, 39]. Ref. [40] provided the model of an H atom that could be trapped in screw dislocations by some diffusion in high temperature growth condition. Besides, hydrogenation lowers the formation energies of point defects including the vacancy and antisite defects [39].

Inverse piezoelectric effect is a result of applied high electric field which adds tensile strain in AlGaN barrier layer due to the lattice mismatch between the AlGaN barrier layer and the GaN buffer. The total strain can exceed beyond crystal elasticity, and consequently crystallographic defects can be created [37]. One simulation of electromechanical stress under the gate at the drain side showed that the maximum electric field opens the possibility for electron injection and inverse piezoelectric effect [41].

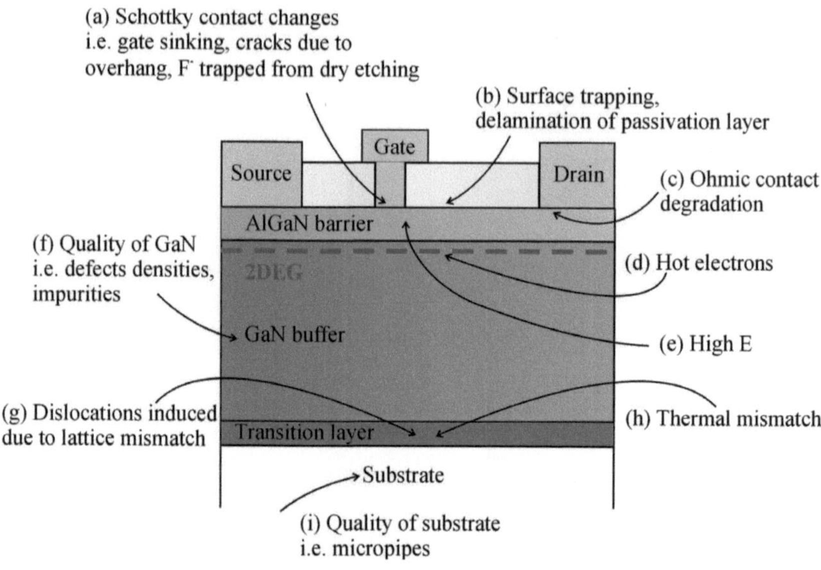

Figure 1.8 AlGaN/GaN HEMTs schematic cross-section, identifying critical areas for device degradation (summarized from Ref. [36, 42]).

Fig. 1.8 depicts critical areas in AlGaN/GaN HEMTs which are related to epitaxial growth quality and process as well. There are possibilities that Schottky and ohmic contacts degrade and imperfect passivation layer are responsible for parasitic charging effects which cause dispersion and lagging effects (see (a), (b), and (c) in Fig. 1.8). One should design carefully Al-GaN/GaN HEMT device concerning the high electric field (~ 6 MV/cm) under the gate at the drain side which can cause degradation due to inverse piezoelectric effect and generate hot electrons (see (d) and (e) in Fig. 1.8). This can lead to crystallographic defects where pre-existing defects (i.e. vacancies, dislocations, impurites) can aggravate. These defect clusters can trap electrons or can become electically conductive and thus reduce device performance. Localized native defects due to material growth condition, and dislocations mostly due to lattice mismatch between substrate and GaN buffer (see (f) and (g) in Fig. 1.8) can act as degradation points. During cooling down after material growth, the wafer may bend which indicates large tensile strain. In a worse case, when thermal mismatch between GaN and substrate is large, even cracks can occur (see (h) in Fig. 1.8). During gate process using plasma etching employing reactive fluorine components, fluorine can be incorporated in the semiconductor and create instabilities there. This step can create crystallographic defects which act as traps under the gate (see (a) in Fig. 1.8). In addition, substrate quality needs careful inspection. This is due to the fact that defects such as micropipes in the substrate can penetrate to the surface and influence the electrical properties of the device (see (i) in Fig. 1.8). Therefore, it is important to discuss both epitaxial growth and device fabrication including defects formation that I will explain in chapter 2.

1.5.1 Heckmann diagram: crystal properties relations

It is necessary to discuss the intrinsic property of GaN as a centrosymmetric crystal. Mechanical, electrical and thermal parameters of centrosymmetric crystal are described by the Heckmann diagram (see Fig. 1.9). There are three "forces" applied to the crystal in the three outer corners: temperature T, electric field E, and mechanical stress σ. Each of these "forces" has direct results: entropy per unit volume S, electric displacement D, and strain ϵ respectively. The three principal effects regarding these pairs (depicted in the thick arrows) [43]:

- In a reversible change, and considering unit volume, an increase of temperature produces a change of entropy $dS = (C/T)\,dT$, where C is the heat capacity per unit volume, and T is the absolute temperature.

- a small change of electric field dE produces a change of electric displacement $dD = \epsilon\,dE$, where ϵ is the permittivity tensor.

- a small change of stress σ produces a change of strain $de = s\,d\sigma$, where s is the elastic compliances.

The three properties, thermal, mechanical and electrical, are correlated to each other such as channel temperature when GaN device is switch on is related to electric field and current distribution. By performing electroluminescence (EL) measurements, Shikegawa $et\ al.$ observed that the EL intensity reveals peaks around the edge of the channel where the electron temperature is high [44]. Theoretically, strain modifies the bandstructure because strain changes the relative positions of atoms in a material and change selection rules for optical transitions [45]. Recently, Ref. [46] showed a correlation between tensile stress and strong luminescence intensity where device under high tensile stress cause a redshift of the peak of photoluminescence intensty.

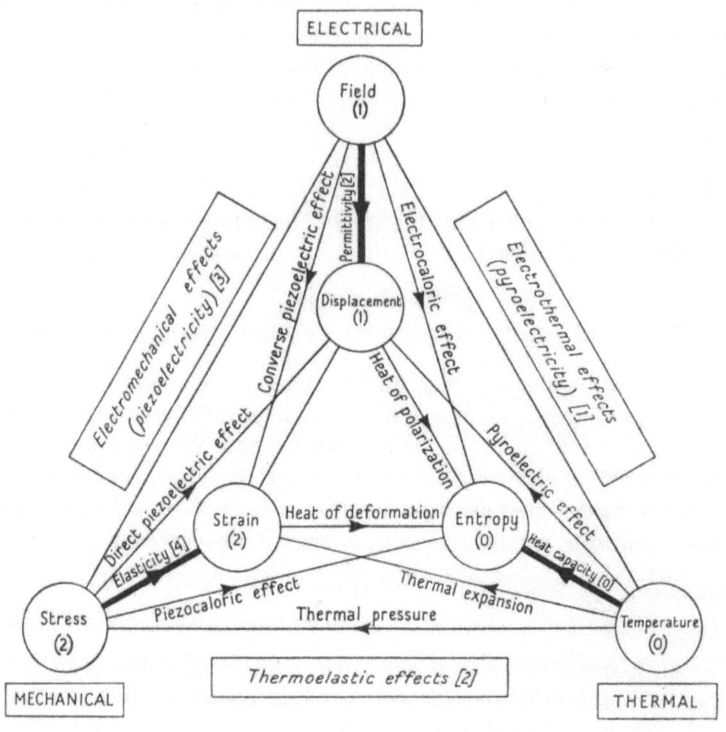

Figure 1.9 Heckmann diagram [43].

1.6 Structures of this thesis

This thesis introduces AlGaN/GaN HEMTs, basic understanding and a brief of reliability issues in chapter 1. Chapter 2 explains AlGaN/GaN HEMTs fabrication from material growth and process technology together with a discussion of potential defect creation. The stressing methods and characterizations including electrical, physical and localized characterizations are discussed in chapter 3. Then, we designed experiments to investigate GaN reliability issues in chapter 4. Results and discussions would be shown in chapter 5 followed by our interpretations of the results in chapter 6. The last chapter provides the conclusions and outlook of AlGaN/GaN HEMTs reliability.

Chapter 2

AlGaN/GaN HEMT device fabrication

2.1 Motivation

AlGaN/GaN HEMTs reliability issues as mentioned in chapter 1 need to be examined carefully concerning the whole technological process chain. Good understanding of device fabrication is crucial to narrow down the degradation problems. The damage that is detected in degraded devices is often due to multiple degradation effects being the consequence of one initial degradation mechanisms. It is therefore very important to stress devices in such a way that the initial degradation effect can be observed without fully damaging the device as a consequence of this effect. This premature degradation detection is not only necessary for intrepreting degradation modes and failure analysis but also important to give fast feedback to device technology improvement. In this chapter, I will explain the fabrication of AlGaN/GaN HEMTs in Ferdinand-Braun-Institut (FBH). After all, it is also essential to discuss defects which can occur during material growth, during process, from the difference material intrinsic properties such as lattice and thermal mismatch, and/or combination of these factors.

2.2 Epitaxy

In this work, the following single crystal line substrates used are: semi insulating (SI) or n-type SiC substrates with 2- and 3-inch diameter. Regular substrate inspections to check defect and mechanical strain distribution across the wafer were performed by cross polarization light microscope (see Fig. 2.1a). Fig. 2.1b shows the X-ray diffraction mapping of the full width

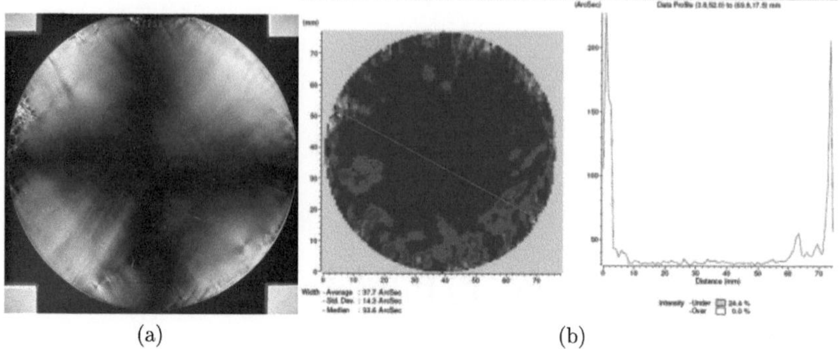

(a)   (b)

Figure 2.1 Substrate routine check: (a) polarized-photo and (b) X-ray mapping.

half maximum of a diffraction peak that characterizes the quality of the crystalline SiC substrate. It can be seen that the crystalline quality of the substrate is much better in the wafer centre whereas at the wafer periphery the crystalline is clearly compromised. To a certain extend this finding also correlated with the crossed polarization imaging of Fig. 2.1a.

Metalorganic vapour phase epitaxy (MOVPE) is the dominant GaN epitaxial growth technique at FBH. It uses metal organic compounds such tri-ethyl or trimethyl-gallium (TEG or TMG) as gallium source and NH_3 as nitrogen source, respectively. The simplified chemical reaction for the GaN growth is the following,

$$Ga(CH_3)_3 + NH_3 = GaN + 3CH_4 + H_2$$

From the equation above, hydrogen involves in GaN deposition. But most of the hydrogen is not coming from this reaction. Hydrogen is the carrier gas for TEG and TMG. Noted that CH_4 at high temperature ($\sim$ 1000 °C) can be decompose. As mentioned before, hydrogenation lowers the formation energies of point defects including the vacancy and antisite defects [39]. Additionaly, hydrogenation could passivate acceptors.

To accomodate the lattice mismatch between SiC substrate and GaN, a thin AlN nucleation layer can be used. The use of an AlN initial layer not only reduces the lattice mismatch between AlN and GaN ($\sim$ 2.4 %) but also promotes surface wetting [47]. The AlN nucleation layer thickness has range of 30-300 nm. The nucleation layer growth conditions influence the GaN nucleation and consequently grain coalescence which then determines GaN dislocation densities, and stress incorporation [47, 48]. Ref. [48] showed that high temperature AlN growth provides a low number of dislocation density

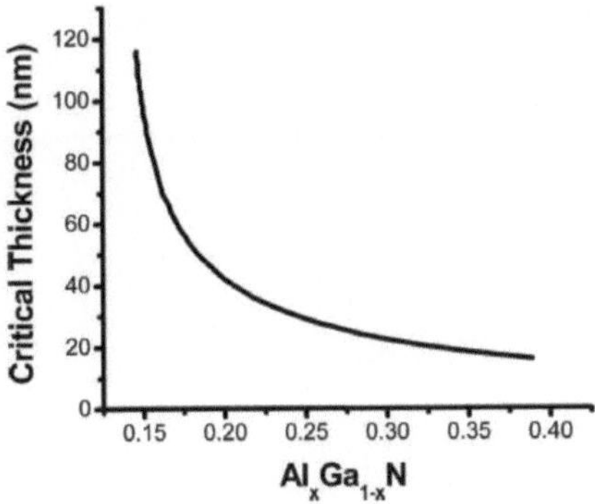

Figure 2.2 Calculation of AlGaN critical thickness on relaxed GaN layer with Fisher model Peierls barrier (dislocation density 10^{10} cm^{-2}) [49].

in GaN buffer layer. AlN thickness study comparison (25 nm vs. 150 nm) showed an order of magnitude of threading dislocations (TDs) difference ($\sim$1x10^9 cm^{-2} vs. $\sim$2x10^8 cm^{-2}) [47].

Thickness of the AlGaN barrier layer on the top of GaN buffer is $\sim$ 25-35 nm as it basically accomodates the lattice constant of the GaN buffer layer. Underneath this layer is heavily tensile strained. This is due to the fact that the Al atoms sit in Ga place which make lattice constant smaller, and the GaN buffer layer forces the system to a GaN lattice constant. Strain in AlGaN increases along with Al concentration. At certain thickness of the AlGaN layer, the elasticity of the AlGaN layer is not strong enough to withstand the tensile stress, and the material starts to relax and to form microscopic cracks. Simulation from Ref. [49] shows that the critical thickness of AlGaN is a function of Al concentration in AlGaN layer (see Fig. 2.2).

A typical epitaxial growth process for HFET structure accompanied by the *in-situ* monitoring of growth temperature and wafer curvature is shown in Fig. 2.3. It shows that wafer acquaintances thermal stress (red line) during epitaxial growth which determines wafer bow curvature (green line). The growth process temperature T_P is measured at the backside of a SiC susceptor. The emissivity corrected the surface temperature of a SiC susceptor T_{true} is measured by a pyrometer with wavelength 950 nm to calculate the growth rate. The *in-situ* surface reflectance data is measured with wavelength 405 nm which is very sensitive to the GaN surface in terms of the coalescence process.

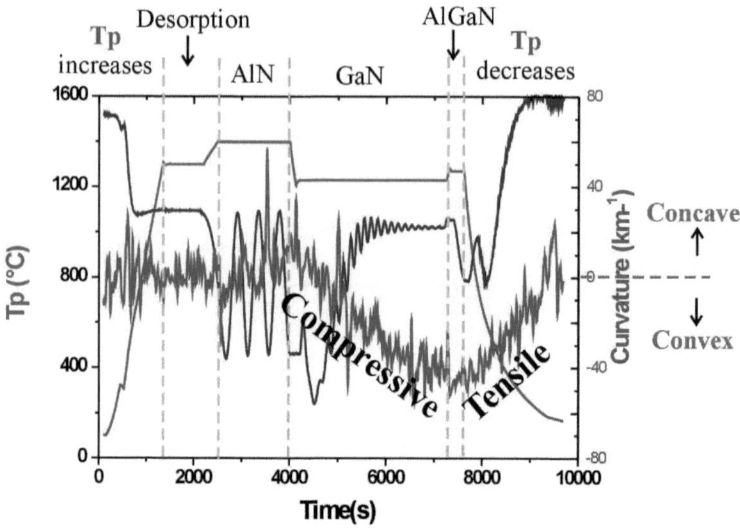

Figure 2.3 Wafer curvature and surface reflectance were monitored during growth using an EpiCurveTT-AR sensor. Red line is the growth process temperature T_P, blue line is the reflectance measured by a pyrometer (450 nm), and green line is the wafer bow curvature. Yellow dashed-lines are marks for temperature changes during epi growth.

2.3 Process Technology

Concerning AlGaN/GaN HEMT device performance and reliability, some optimizations of technology in critical areas are developed as depicted in Fig. 2.4.b. T-gate structures with a longer wing at the drain side acting as a field plate (FP) [50, 51] are applied to reduce high electric field under the gate at the drain side. A new embedded gate technology has been developed in which the nitride is deposited at an early stage of processing and gates are defined by nitride etching and subsequent metallic gate definition. Low resistance ohmic contacts R_C (< 0.5 Ohm cm) are essential for efficient high power electronic device operation. The metal diffusion because thermal annealing leads to rough surface morphology and difficult to make good line edge definition of ohmic contacts. The rough surface morphology can be a problem for homogeneous issue of active areas (will be shown in electroluminescence measurements in chapter 5). The problem of good line edge definition does not facilitate low S-G distance for high speed device operation, and hence it deteriotes the performance of high speed HEMTs [52].

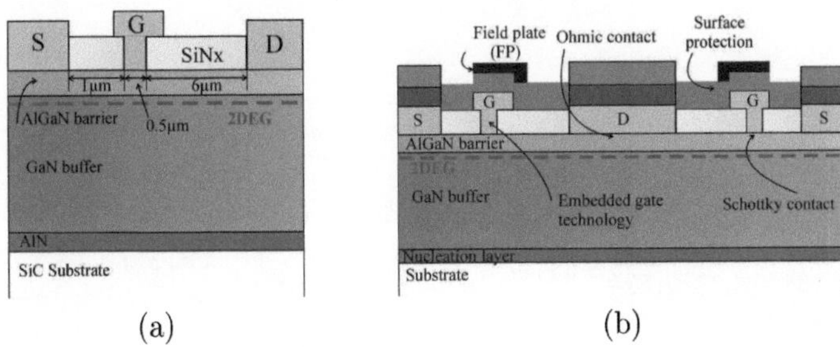

(a)   (b)

Figure 2.4 AlGaN/GaN HEMT (a) cross-sectional dimension of 2x125 μm device used for reliability measurements, and (b) optimizations of process technology for critical areas.

Typical metallic compositions for ohmic contacts are: Ti/Al/Ni/Au, or Ti/Al/Mo/Au or Ti/Al/Ti/Au/WSiN$_x$. The ohmic metal contacts are formed by rapid thermal annealing (RTA) around 800 °C. If ohmic contacts are formed in n-GaN layer, during annealing, Ti reacts with GaN and forms TiN. This reaction extracts N from GaN and generates N-vacancies in GaN layer. Additionally Ga is dissolved in Mo and Au which leads to further inter-metallic compound/semiconductor interaction and thus helps to form a good ohmic contact. These vacancies acts as n-type dopant and create a highly doped region in the vicinity of the interface, and generates the foundation for tunneling contact mechanism. If ohmic contacts are formed in AlGaN layer, it has been observed that the contact resistivities of Ti/Al metaliza-tions on AlGaN/GaN increased with an increasing Al concentration (due to an incrase of band gap of AlGaN) and with an increasing AlGaN thickness due to the increasing tunneling depth. However it has been observed that TiN protrusions which are formed along dislocations, penetrate through the AlGaN barrier layer, and hence may directly contact the 2DEG [53].

Regularly checked with visual inspection such SEM as depicted in Fig. 2.5. (a) reveals the edge delineation problem which causes a problem to place gate closer to source than 0.5 μm (might create short circuit), and fence-like WSiN structures that might compromise yields. Surface passivation is intended to reduce surface traps and avoid current collapse. However, non-continous surface passivation i.e. a trench opening to form the gate and subsequent SiN$_X$ passivation above the metallic gate introduces very complex compressive and tensile stress situations in the gate area which may give rise to premature degradation effects in these areas (see Fig. 2.5. (b)). Ref.

[54] observed other optimization such as second passivation and different encapsulation of passivation have an impact to gate leakage.

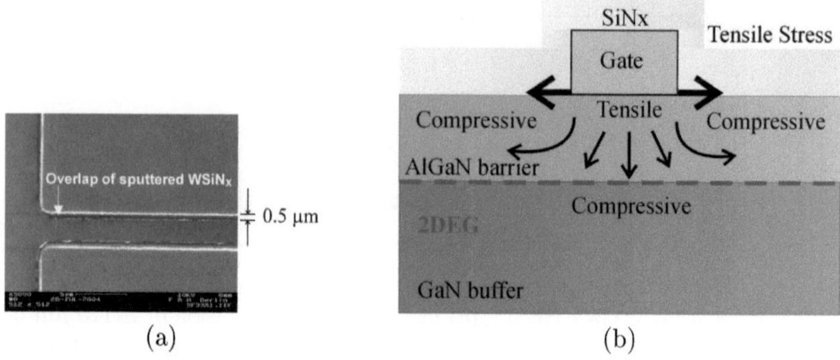

(a)　　　　　　　　　　　　(b)

Figure 2.5 Examples of process problems (a) SEM picture of Ti/Al/Ti/Au/WSiN$_x$ ohmic contact with edge delineation [52], and (b) schematic of area below the gate of AlGaN/GaN HEMT with a non-uniform SiN$_x$ passivation layer induces compressive and tensile stress [49].

2.4 Defects

It is very crucial to discuss defects in this chapter because they strongly determine electrical, optical and thermal properties of semiconductors It has been suggested for the first time by Ilegems and Montgomery that native defects dominate the transport properties of bulk GaN [5] explaining the n-type conductivity of bulk GaN. In this section, I discuss types of defects which affect AlGaN/GaN HEMT device performances and reliability.

Types of defects

Based on dimension, defects in solids can be categorized as the following (see Fig. 2.6):

- 0D: point defects, i.e. vacancies, self-interstitial atoms, substitutional impurity atoms, interstitial impurity atoms and antisite defects.

- 1D: dislocations, i.e. edge and screw dislocations.

- 2D: grain boundaries i.e. tilt- and twist-type, stacking faults, and external surface.

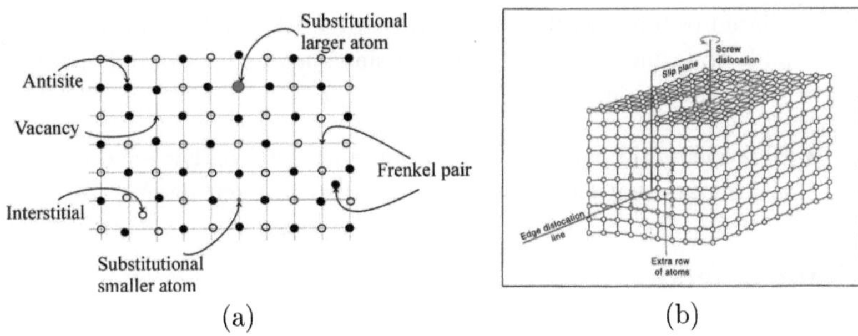

(a)   (b)

Figure 2.6 Some examples of defects (a) point defects (grid lines are a guide for eye) (b) edge and screw dislocations [55]

- 3D: bulk or volume defects, i.e. voids, cracks and foreign inclusions.

Defects formation

The formation probability of crystal line defects depend on:

- Crystal structure and unit cell atoms. GaN has a relatively low atomic packing fraction of 0.42 [56]. This represents a moderate degree of openness in the unit cell but also depicts a large differences in radii (Ga has a large and N has a small covalent radii). This makes the formation of interstitial and antisite defects unfavorable; only vacancies have formation energies that are low enough. This fact is important in interpreting degradation mechanisms.

- Lack of stoichiometry. This influences the formation energies of defects in the growth process, such as Ga-rich growth condition leads to a smoother surface. A reduced growth temperature leads to C-incorporation in the grown layers (in this case C stems from the metalorganic precursors).

- Growth conditions. Defects can enter a material along with the diffusion of wanted, or unwanted, impurity atoms such as hydrogen during crystal growth which can lower the energy formation of point defects [39]. These unwanted materials are dependent on growth conditions [57].

- Mechanical property. Defects can also be introduced into materials by the processes of plastic deformation. This mechanical property is

19

related to temperature, lattice mismatch, thermal expansion mismatch and strain incorporated by processing such as for example passivation or metallization.

- High energetic particles interaction with lattice. For example, plasma etching and hot electrons can cause displacement damage [35, 36, 58].

Point defects

Among GaN native point defects, vacancies have the lowest fomation energies [7]. Under p-type conditons the nitrogen vacancy is dominant, and under n-type conditions the gallium vacancy is dominant [59, 60]. Source of unitentional n-conductivity of GaN are oxygen- and silicon incorporated for certain reactor types [7]. The defects are not independent from each other: they are coupled via the condition of charge neutrality. The total sum of electrons in the conduction band, holes in the valence band and charges trapped in the defects must be zero. Consequently, when a potential difference is applied in GaN-based devices (nonequilibrium condition i.e. OFF-state), charges trapped in the defects can be released and may degrade the device.

A low certain formation energy is required to create defects in larger concentration under conditions of thermodynamic equilibrium. For all growth temperatures the gallium vacancies are those point defects with the highest concentration even at high temperature of 1300 K (growth temperature) [59].

Point defects i.e. V_{Ga} and its complexes with one or more O_N have very low formation energies at different positions near the threading-edge dislocations [61]. If the point defects in are mobile, a number of reactions can occur i.e. an interstitial can annihilate with its own or another vacancy. Interstitials or vacancies can cluster, agglomerate and trap impurity atoms. Mobile point defects are known in other semiconductors i.e. two types of mobile point defects in Ge surface were observed by STM at temperatures below 80 °C [62]. Ref. [63] pointed out that the mobile defects can be trapped by impurity atoms, thus forming impurity-related complexes. Limpijumnong and Van de Walle investigated the diffusion of relevant native point defects in wurtzite GaN crystals. Gallium interstitials Ga_i migrate via an interstitialcy mechanism with an unexpectedly low barrier of 0.9 eV. They are mobile at temperature slightly below room temperature. The migration barrier for gallium vacancy V_{Ga} is significantly low of 1.9 eV. For all these defects the lowest-energy migration path results in motion both parallel and perpendicular to the c-axis [64]. This should be noted for the interpretation of lifetime effects.

Luminescence is a powerful method to detect point defects. Reschchikov and Morkoc studied luminescence of point defects for different states in energy band gap in doped and undoped GaN [60, 65] (see Fig. 2.7). As several other authors, they also pointed out that $V_{Ga}-O_N$ defect complexes i.e. $V_{Ga}-O_N$ contribute to well known broad band yellow luminescence (YL) besides other defects i.e. dislocations at low-angle grain boundaries [66].

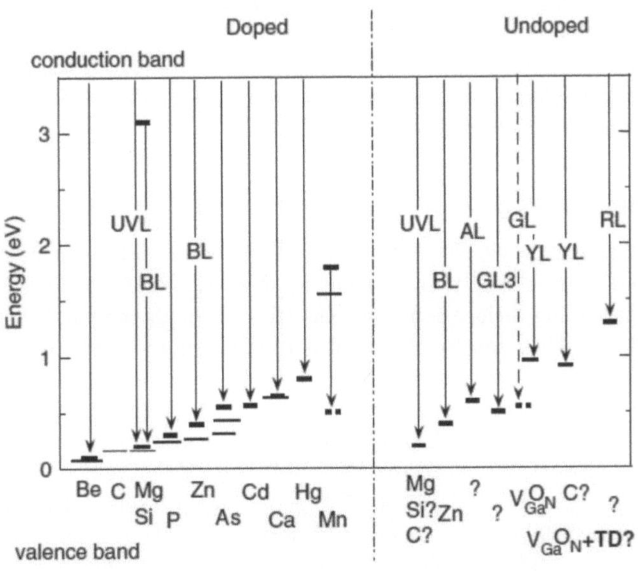

Figure 2.7 Radiative transitions associated with major doping impurities [65].

Dislocations

Abundant studies of dislocations in GaN are available. It is well known that heteroepitaxial growth of GaN results in formation of high threading dislocation (TD) densities. The lowest reported TD densities in GaN are $\sim 10^6$ cm^2. TDs are problematic: they can be sources of nonradiative recombination [67, 68, 69] and modify crystal potentials creating localized electrostatic and strain fields [69, 70, 71, 72, 73]. TDs also scatter carriers and reduce the carrier mobility [74] as well act as pathways for current leakage [75, 76, 77] and therefore can lower the breakdown voltage V_{BR} [78]. Consequently, the fabrication of GaN layers with low TD density remains one of the most important challenges in the GaN epitaxy. The lowest threading dislocation (TD) densities are reported to be in the range of $\sim 10^6$ cm^{-2} [79]. Screw and mixed dislocations dominantly contribute to nonradiative recombination compared to edge dislocations [68, 80]. They create states in GaN band gap which can accept or donate electrons, and may become charged in doped material [40]. The stress field produced by the dislocations can accumulate electrically active impurities and point defects near the dislocations [69]. Ref. [81] correlated dislocations to the existence of the bright spot during electroluminescence measurements at OFF-state. Meanwhile Ref. [82] associated bright spots during electroluminescence measurements at OFF-state with electron release from traps of defect complexes (i.e. V_{Ga}-O_N).

Surface defects

As mentioned previously, lattice mismatch between the substrate and the GaN buffer layer results in dislocation formation in GaN layer. These, so-called threading dislocations can penetrate through the whole GaN/AlGaN layer stack. They can be observed as pits on the (Al)GaN surface by AFM measurements. These pits are the terminations of dislocation lines and contribute to the surface roughness. Heying *at al.* found the GaN surfaces to be dominated by three dislocation mediated structures: pinned steps, spiral hillocks and surface depressions [83]. Ref. [84, 85] pointed out that the origin of the hillocks on N-polar c-plane and m-plane are screw dislocations. The association of hillocks to dislocations was directly shown by AFM measurements that reveals the spiral pattern of atomic steps [86].

Defects originating from SiC substrate micropipes can run through the buffer and up to the surface of AlGaN buffer which influence device characteristics such as IV-output, transfer and gate leakage [87]. Bang *et al.* correlated wafer yield to various size and densities of the hollow of these structures. It is known that they introduce free carrier densities close to the active region and therefore influence 2DEG. Similar effects are applicable to Si substrates [88], defects in any case influence the AlGaN/GaN HEMTs device performance. Depending on distance from active area, deep pits may influence pinch-off voltage V_{TH}, maximum drain current I_{Dmax} and breakdown voltage V_{BR} [89] (see Fig. 2.8). These pits are orginating from the substrate and are present until the top of AlGaN surface layer irrespective of buffer technology. Deep pits can be the original sources of high leakage current through the buffer and substrate which yield devices with low V_{BR}. The prevention and control of deep pits largely depends on the epitaxial growth technology [90].

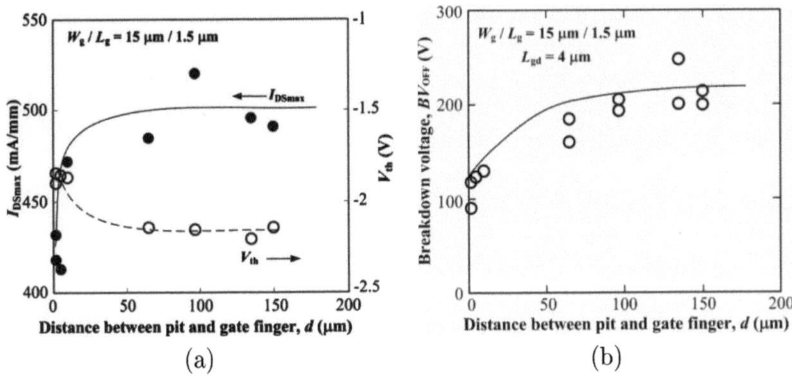

Figure 2.8 AlGaN/GaN HEMTs electrical characteristics in dependence from pit distance from gate finger (a) I_{Dmax} and pinch-off voltage V_{TH}, and (b) breakdown voltage V_{BR} [89].

Those defects discussed before may also affect GaN optical properties which can be observed at certain wavelenghts such as yellow luminescence; however, usually a large amount of damage is needed for detection. In contrast, electrical properties are usually affected already by small quantities of defects which introduce levels in the forbidden gap. The electrical properties include carrier concentration is usually sensitive to lattice defects. Ref. [74] examined 2DEG transport in the AlGaN/GaN HEMTs is mostly dominated by dislocation scattering with a typical dislocation density above 5×10^7 cm^{-2} for Al contents ranging from 8-35 %. They predicted the mobility of the 2DEG to exceed 2000 cm^2 V^{-1} s^{-1} as the dislocation density falls below $\sim$ 5×10^7 cm^{-2}.

Chapter 3

Device stressing and characterizations

For systematic investigations of AlGaN/GaN HEMTs reliability problems, one should refer to basic properties of nonsymmetric materials as described in Heckmann diagram (see subsection 1.5.1) . It relates electrical, mechanical and thermal properties to each other and is therefore well suited for a formal description of the dependencies that might occur during device stressing. It could for example relate thermal, mechanical, electrical and optical properties to each other and thus provide the basis for a more detailed understanding of degradation effects. In this work we have concentrated on detecting degradation thresholds, which mean that we have to monitor the device behavior once a degradation of instability effect occurs. Therefore, we have established DC-Step-Stress test to find out degradation thresholds which are also intended to provide a short term information on the principal robustness of the devices. Indeed, these stresses deliver meaningful fingerprints and are able to detect characteristic properties in dependence on device material properties, technology and device layout. The stressings are accompanied by characterizations such as IV-measurements, electroluminescence and localized characterizations. Localized characterizations include cross-sectioning by focused ion beam (FIB) and mechanical grinding for transmission electron microscopy (TEM) and energy dispersive x-ray spectroscopy (EDX) investigations. The devices under test have size of 2x125 μm with FBH standard source-gate space $L_{SG} = 1$ μm, gate-drain space $L_{GD} = 6$ μm, gate length $L_G = 0.5$ μm, unless otherwise specified.

3.1 Robustness tests

DC-Step-Stress tests consist of 5 V step ramping of drain-source voltage V_{DS} every two hours on pinch-off transistors ($V_{GS} = -7$ V). They are performed at room temperature in darkness (see Fig. 3.1). Since this electrical stressing is performed at OFF-state, one can see the effect of applying high electric field without thermal stress (channel is not open). The irreversible evolution of one or both, gate leakage current I_G and subthreshold drain current I_D are observed during stressing and have been taken as a criterion for the onset of device degradation so called critical voltage V_{CR} when one or both of those currents increase (see Fig. 3.2). In the several steps before the point of degradation, trap related chargings are observed. The number of devices for our screening standard DC-stressing on wafer is a minimum five devices across the wafer: one from the centre, two from the edge and two from mid-centre. This insures a certain statistical relevance.

Definitely, one cannot compare the degradation mechanism at OFF-state with degradation mechanism at ON-state when the channel is open. It is likely, that the degradation mechanisms are different between OFF- and ON-states [35]. DC-stressing at OFF-state pronounces the impact of high electric field on devices meanwhile at ON-state. It is therefore a good method to separate this from current assisted degradation effects. However, one should consider the piezoelectric effect for both ON- and OFF-state stress conditions.

Another test so called the high temperature reverse bias (HTRB) applies both, thermal and electrical stressings simultaneously to the device. HTRB is triggering field assisted and thermally activated degradation mechanism simultaneously. Therefore, one should carefully intepret the results because it is not easy to distinguish how each stress type affects the device performance; temperature or high electric field and/or which one is the trigger to the other.

3.2 Long term lifetime tests

Typical wear-out times of semiconductor devices are in the order of 1×10^6h ($\sim$100 years or more) hence accelerated testing conditions are needed. Many chemical processes associated to device failure progress exponentially with temperature i.e. diffusion and metal migration. Commonly, an accelerating test by applying thermally activated stress on a number of devices is a standard procedure to predict median time to failure (MTTF). The Arrhenius model for lifetime prediction is possible if degradation mechanisms follow an exponential law with temperature

$$\text{MTTF} = A \exp (E_a/kT)$$

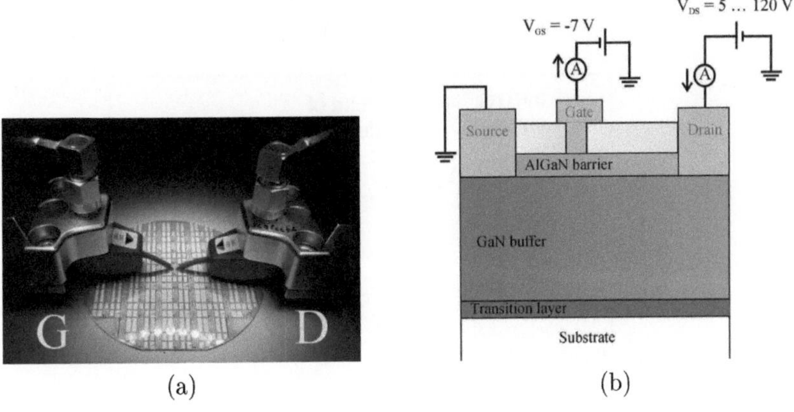

(a)   (b)

Figure 3.1 DC-Step-Stress Test on wafer with (a) two coplanar probes connected to gate voltage source (G) and drain voltage source (D). The probe has three fingers where the middle finger is connected to gate or drain voltage source (signal), and the two fingers are connected to source (ground), and (b) schema of electrical circuit for DC-Step-Stress Testing.

where E_a is the activation energy associated to the particular failure mechanism, T is the device temperature, k is Boltzmann's constant and A is a constant.

Common model to analyse the life test data which involves fitting the failure rate at a particular temperature is Weibull distribution. MTTF is measured by extrapolating a number temperature points taken at high temperature to the temperature of device application. The number of data points at each temperature should be in good confidence level which increases with an increasing number of samples. A lognormal plot of the failure times that would be used to originate MTTF graph based on Arrhenius formula, ideally lines up the data points in a straight line. Lifetime prediction to a specific operation temperature is possible be extrapolating the resulting straight line to the given temperature. This is possible only if the following are fulfilled: 1) the failure mechanisms are identical at all high temperature test points, and 2) the failure mechanism at high temperature is the same at the device application temperature [29], and 3) the failure mechanisms exponentially depend on temperature. If the slope at each temperature is similar, a single failure mechanism (assumption 1) can be considered. One should be careful with assumption 2 since the accelerated high temperature tests do not detect the existence of a low activation energy failure mechanisms which may lead to highly overestimated MTTF values (see Fig. 3.3). Failure mechanisms

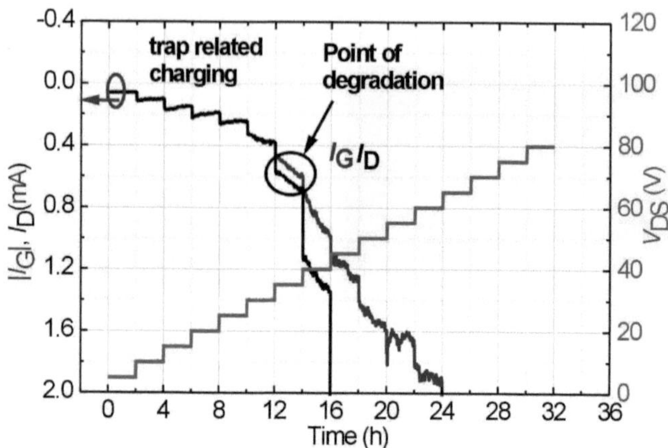

Figure 3.2 An example of DC-Step-Stress test result. The gate leakage, subthreshold drain current and drain-source voltage V_{DS} are represented by blue, black and red lines, respectively. Trap related chargings are observed during single steps before gate leakage and/or subthreshold drain current finally start to degrade irreversibly. Critical voltage V_{CR} of this device is determined to 35 V. This is the drain voltage V_{DS} where degradation effects are clearly seen for the first time.

which are or only weakly activated by temperature are possible i.e. high electric field causes inverse piezoelectric effect [37].

The channel temperature used for MTTF measurements has been determined by several methods as shown in Table 3.1 with each advantage and limitation.

3.3 Electrical characterizations

Static and dynamic IV-characterizations were performed on AlGaN/GaN HEMTs before and after DC-stress at base temperature 16 °C altogether with output transfer and diode characteristics. In this study, generally static IV-characteristics are measured by increasing drain voltage V_{DS} (from 0 to 30 V) for each value of gate voltage V_{GS}(from -7 V to +1 V) with a specific power dissipation limit P_{diss} (3 W/mm). Transfer curve characteristics are measured at drain voltage $V_{DS} = 10$ V from gate voltage V_{GS} -7 V to

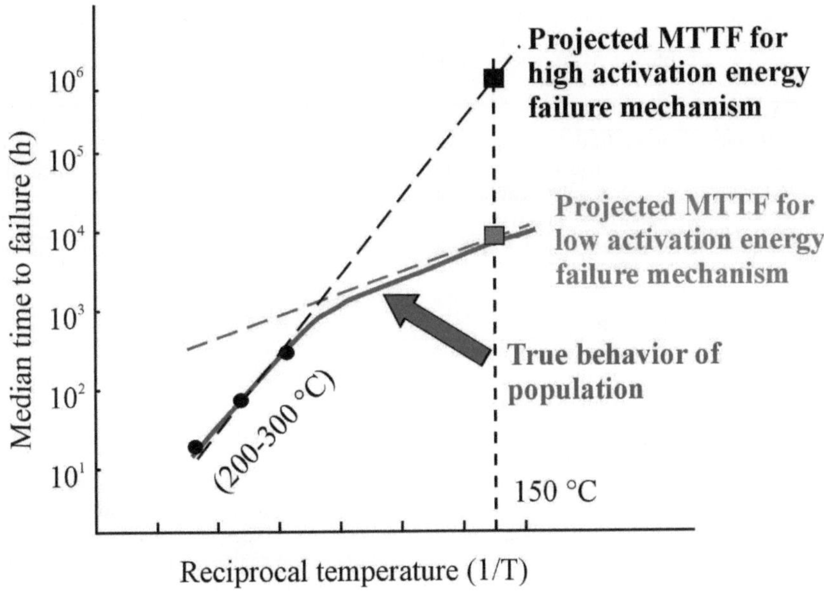

Figure 3.3 MTTF estimation illustration of overestimated MTTF at high accelerated high temperature test conditions to the actual device application temperature [29].

+1 V. Diode characteristics are measured in positive direction from gate to drain and to source $V_G = 0$ to 2 V with maximum current of 10 mA/mm. And for negative direction, they are measured from source and drain to gate contact with negative voltage from -1 to -10 V with negative current limit of 1 mA/mm.

When the static IV-output characteristics are performed at comparably long scan times in such a long dwell time for example 1 V/s from 0 to 30 V of V_{DS}, these conditions may give enough time for electrons to change trapping states in the surface or AlGaN/GaN buffer interface. This trapping effect is observed as kinks at knee voltage usually below 10 V V_{DS} (see red arrow in Fig. 3.4a). In this figure also self heating of the device can be oberved by a decay of drain current at high power dissipation levels. This is due to a reduced charge mobility in the channel due to phonon scaterring. This causes a drop of saturated current (pointed by blue arrows in Fig. 3.4a).

Dynamic IV-characteristics are performed on a device by pulsing from a static quiescent point to the necessary locations throughout the IV plane

Table 3.1 Methods to measure channel temperature device [29]

Method	Technique	Limitations	Advantages
IR imaging	using IR detector to determine surface temperature which requires calibration of surface emissivity	resolution $\sim$ 3-5 μm	Fast, cover large areas, die-attach integrity
Liquid crystal	coating liquid crystal, power up slowly and observe transition temperature	invasive and range of transition temperature is limited	No calibration required
Electrical	applying constant current to calibrate forward-biased diode. temperature is proportional to forward voltage	indirect measurement	in situ monitor of circuit temperature
Raman spectroscopy	observing the raman spectrum shift	slow and for best resolution needs large integration time	need a model to determine peak channel temperature, high resolution

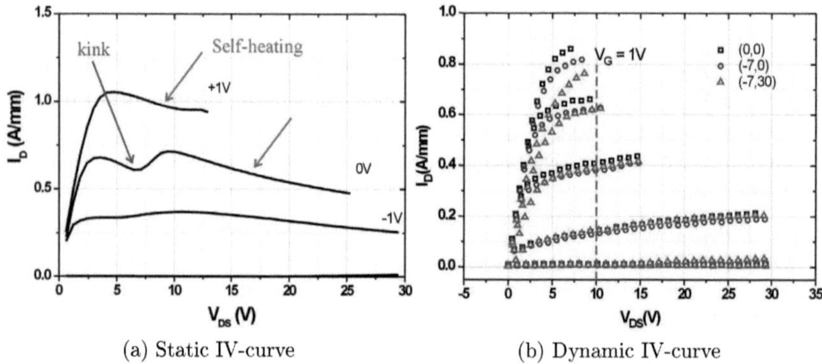

(a) Static IV-curve (b) Dynamic IV-curve

Figure 3.4 An example of (a) static IV-curve with kink at knee voltage due to trapping effect, and drop of saturated drain currents due to self-heating, and (b) pulsed IV-curve with each bias point (0,0), (-7,0) and (-7,30) are represented by open black squares, blue circles, and red triangles respectively.

(see an example of pulsed IV-curve in Fig. 3.4b). The pulse length is short enough to provide an IV-measurements isothermal (quiescent thermal) and isodynamic (quiescent trapping) conditions [91]. Pulsed I/V measurements are especially suitable to characterize the influence of trapping effects on the electrical performance of devices at well defined bias point conditions. These measurements rely on the fact that trap states usually have large time constants. If a device is then constantly biased at a certain point and the IV characteristics is measured starting from and returning back to this bias point within very short times (less than 1 μs) and residing at the bias point relatively long (large duty cicle) the trap situation at this point can be considered as constant- the trap are not able to respond to the short measurement pulses. Hence the IV characteristics measured exactly reflects the internal charging situation of the device in this particular bias point.

To characterize device trapping effects the following three bias points are particular interest:

- If the device is biased at both, drain and gate voltage equal to 0 V, practically no internal electrical fields are present in the device. Thus this situation practically reflects the situation of a non trapped device.

- If the device is biased at zero drain voltage and at pinch-off gate conditions, electrical fields are dominating in the gate area only. Hence traps related to a limited volume around the gate are affected. This condition is referred to as "gate lag".

- If the device is biased OFF-state, however at a higher drain voltage, the gate high field region extends more and more towards the drain and to the buffer which means, that now traps in these regions are additionally affected. This condition is referred to as "drain lag".

As a consequence, pulsed I/V measurements prove informations on the basic location of trap states within the device. In this study, the pulse duration is 200 ns with a separation time of 0.5 ms and the quiescent points are (V_{GS}, V_{DS}): (0,0), (-7,0) and (-7,30). A gate lagging factor z_G is defined which rates the nearly trap free pulse measurement at zero bias conditions to the actual measurement at a given bias point. The z_G is defined as

$$z_G = (I_{D(-7,0)} - I_{D(0,0)})/I_{D(0,0)}$$

for gate lag and

$$z_D = (I_{D(-7,30)} - I_{D(0,0)})/I_{D(0,0)}$$

for drain lag conditions where $I_{D(0,0)}$, $I_{D(-7,0)}$ and $I_{D(-7,30)}$ are the drain current at (0,0), (-7,0) and (-7,30) respectively. The values are usually taken at constant drain bias V_{DS} of 10 V and a gate bias V_{GS} of $+1$V to ensure comparability (see green dashed line in Fig. 3.4b).

3.4   Optical characterizations

Early GaN luminescence measurements were performed by electron-beam and optical excitation [3, 92]. Any spontaneous light emission from an excited state to a lower state is called luminescence. The main requirement for emission is that the system is brought out of equilibrium conditions which needs some form of excitation. Excitation by an electric current triggers electroluminescence (EL), by optical excitation produces photoluminescence (PL), and by an electron beam causes cathodoluminescence (CL) [93]. Luminescence is a very strong tool for detection and identification of point defects in semiconductors, especially in wide-band-gap varieties [60]. It is also a non-destructive test and a sensitive measurement. PL and CL measurements are commonly used to characterize GaN after epitaxy growth.

Electroluminescence (EL) measurements are known as a useful tool to localize potentially defective regions in the device topology [36]. Electroluminescent emission has initially been studied in GaAs and InP based heterostructure field effect transistors (HFETs) in order to investigate hot electron induced breakdown [94], impact ionization [95] and the conduction-to-conduction band (intra-band) transition of electrons [96]. In AlGaN/GaN HFETs technology EL has also been claimed as a valuable tool to provide a deeper insight in electronic properties of devices. It has been pointed out that the EL is possibly due to hot electron [36, 97] and intra-band transitions [98]. EL emission has therefore been proposed as a way to probe the regions of high electric field in a device.

A typical EL image at $V_{DS} = 10$ V and $V_{GS} = $ -1 V (ON-state) with integration time is 1 s is shown in Fig. 3.5. EL emission at ON-state condition is generated in the active areas between gate-drain contacts and relatively homogeneous.

Typical normalized EL intensity of AlGaN/GaN HEMTs vs. transfer characteristics is shown in Fig. 3.6. The EL intensity curve starts when the 2DEG channel begins to open and increases abruptly. The peak of EL intensity is located at about a quarter of maximum drain current. With respect to the spectral distribtuion of the EL light, Ref. [98] observed a broad peak at around 670 nm ($\sim$ 1.85 eV) of AlGaN/GaN HEMTs EL spectrum. This EL spectrum is different from unbiased PL spectra, however impact ionization is unlikely the source of EL in AlGaN/GaN HEMTs.

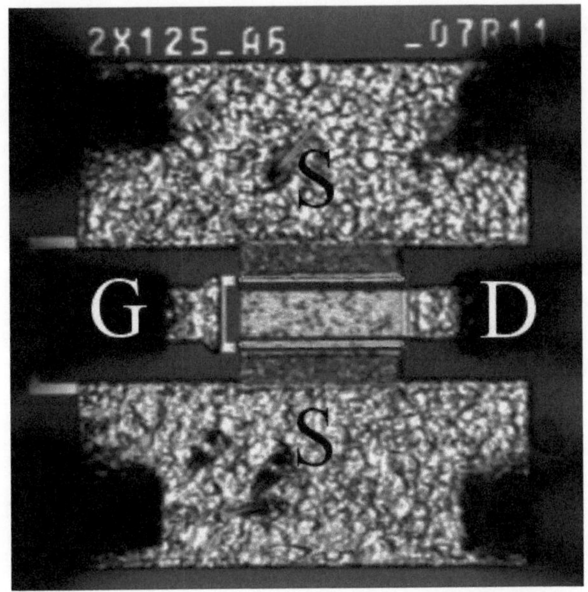

Figure 3.5 Superimposed emission and optical image of AlGaN/GaN HEMT 2x125 μm device at V_{DS} = 10 V and V_{GS} = -1 V (ON-state).

Fig. 3.7 shows our simulation of electric fields of AlGaN/GaN HEMT device in the channel during transfer characteristics at three V_{GS} bias points (marked in Fig. 3.6). This electric field simulation explains the EL intensity evolution vs. transfer characteristics in Fig. 3.6 as the following: when 2DEG channel is pinched-off (V_{GS} = -5 V), the electric peak reachs the maximum value of 0.85 MV/cm under the gate in the drain side. The electric field decreases slightly ($\sim$ 0.6 MV/cm) when the 2DEG channel is open and drops further about 0.4 MV/cm when gate voltage increases (V_{GS} = +1 V). Kolnik *et al.* calculated the threshold of electric field for electrons to occupy the second conduction band in Γ direction of wurtzite GaN bulk is to 0.3 MV/cm [99]. The electric field simulation results in an electric field intensity of typically of 0.6 MV/cm at ON-state which more than enough to excite electrons in the channel occupying the second conduction band in Γ direction.

When the gate-source voltage V_{GS} further increases i.e. V_{GS} = +1 V, the drain current I_D increases as well. This results in a more homogeneous distribution of the potential along the channel hence the electric field peak

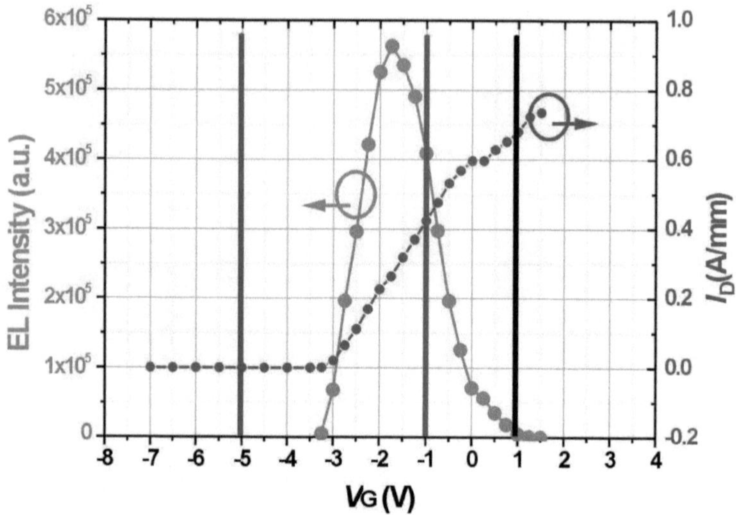

Figure 3.6 The typical EL intensity vs. transfer characteristics at $V_{DS} = 10$ V shows a bell-shaped curve EL with its peak around a quarter of the maximum drain current I_D. The red, light blue and black lines are electric field simulation points in the 2DEG channel at V_{GS} = -5 V, -1 V and +1 V respectively (see to Fig. 3.7).

at the drain edge of the gate is reduced. This is described as electric field decrease, and explains drop of the EL intensity at high gate-source voltage i.e. $V_{GS} = +1$ V. Ref. [99] calculated the difference between first and second conduction band in WZ GaN bulk at Γ direction is about 2.2 eV. Therefore, it is plausible to explain the peak of the EL spectrum at 670 nm as intra-band transition [98] as depicted in Fig. 3.8.

We have frequently performed EL measurements to characterize GaN devices before and after stressing. EL measurements in this work have been performed at the Technical University Berlin using a Hamamatsu Photon Emission Microscope (Phemos 1000) in the darkness. The system is equipped with a liquid nitrogen cooled Si-CCD detector (- 50 °C) with spectral sensitivity in the visible and near infrared regime. The EL measurements were performed along the transfer characteristics. The total EL intensity is counted from emission distribution over the two finger devices.

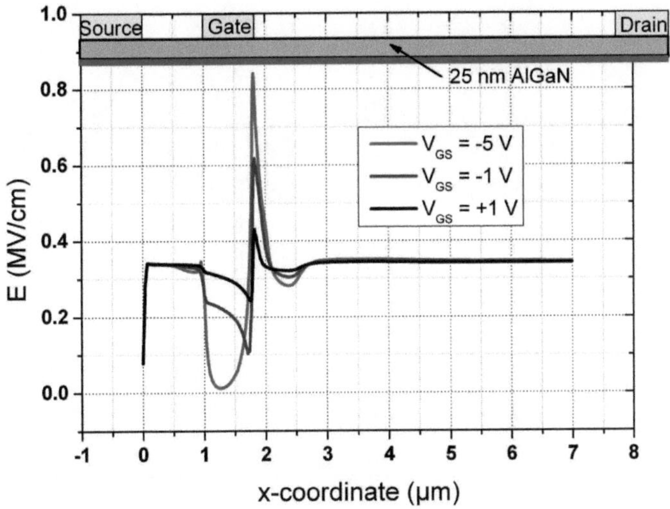

Figure 3.7 Simulation of absolute intensity of electric field of AlGaN/GaN HEMT device with 25 nm thickness of AlGaN layer and 0.7 µm gate length. The electric fields are calculated in the channel (26 nm depth from the surface along the green line) with bias conditions: V_{DS} = 10 V and V_{GS} = -5 V (red line), -1 V (blue line) and +1 V (black line) [Courtesy of E. Bahat-Treidel].

3.5   Localized structural analysis

Cross-sectioning methods are required to investigate the structural built-up and possible structural and compositional changes in devices. We applied two commonly used methods to obtain a device cross-section: mechanical grinding followed by ion milling and focused ion beam (FIB) technique. The obtained samples were subsequently analysed using transmission electron microscopy (TEM). Compared to FIB preparation the mechanical grinding method is extremely time consuming. Furthermore, using this conventional preparation method it is often impossible to prepare strongly localized small device areas. On the other hand this method introduces a less number of artifacts into the prepared cross-section, whereas FIB preparation uses bombarding by Ga^+ ions and often results in Ga ion implantation, strong surface amorphization and even formation of liquid Ga droplets on the lamellae sur-

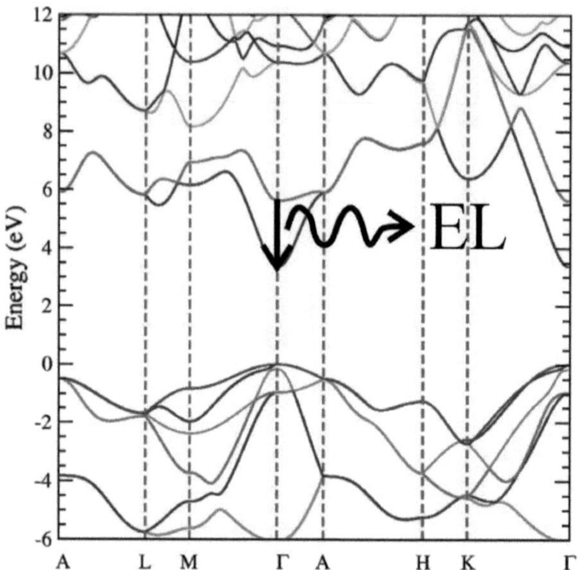

Figure 3.8 Schematic of EL explanation from intra-band transition [98]. Electrons are being excited by high electric field and occupying the second conduction band in Γ direction. These electrons can emit photon during transition to the first conduction band. WZ GaN energy band structure is after [100].

face. However, FIB preparation has a good advantage that it can be used to cut a localized area of interest to see the cross-section of device structures. To minimize the unintentional damage caused by the FIB preparation, one should protect the localized areas of interest using e.g.Pt coating (see the difference of FIB cross-sectioning with and without Pt coating in Fig. 3.9).

FIB preparation conditions for lamella:

- 2 μm thick Pt stripe deposition of area 25 μm x 2 μm, the current is 1000 pA, filled box with dwell time of 0.2 s was performed in 3 min.

- Regular cross section (RCS) for area of 25 μm x 15 μm, the current is 2700 pA, dwell time of 1 s, was performed in 40 min. Estimated depth is about 8 μm.

- Cleaning cross section (CCS) for area of 25 μm x 1.4 μm, the current is 1000 pA, dwell time of 1 s, was performed 20 min. The sample

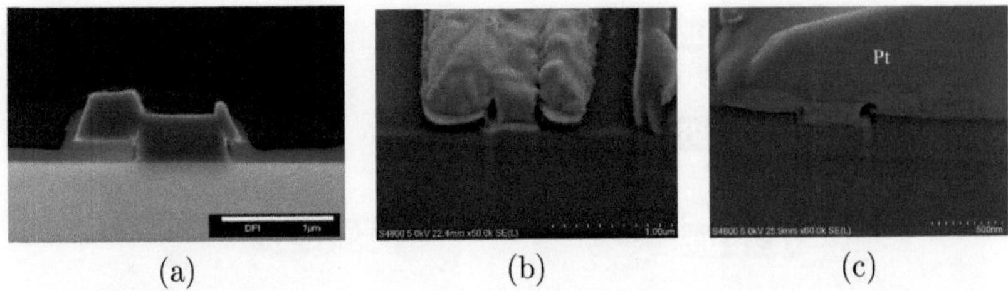

(a)   (b)   (c)

Figure 3.9 Cross-section of gate area of AlGaN/GaN HEMT devices obtained by (a) mechanical grinding (notice T-gate wings are not symmetric), (b) by FIB without Pt coating and (c) by FIB with the protecting Pt coating. Without Pt Ga ion used in FIB cause a roughening of the gate metal wall (b). Note: (b) and (c) are the cross-sectioned T-gate in the same device [Courtesy of P. Kotara].

is rotated by 180°, then RCS and CCS were performed with the same conditions above. The lamella thickness is about 700 nm. Then sample was tilted by 45° for U-cut.

- Filled box was determined for area of 18 μm x 0.7 mm, the current is 350 pA in 15 min. The sample is rotated again by 180°.

- Filled box was determined for area of 18 μm x 0.7 mm, the current is 350 pA in 15 min. The sample is tilted back for window cut.

- CCS for area of 7 μm x 0.6 μm with the current of 350 pA, dwell time of 1 s, was performed in 2 min.

- CCS for area of 7 μm x 0.4 μm with the current of 150 pA, dwell time of 1 s, was performed in 2 min. The sample is rotated by 180°.

- CCS for area of 7 μm x 0.6 μm with the current of 350 pA, dwell time of 1 s, was performed in 2 min.

- CCS for area of 7 μm x 0.4μm with the current of 150 pA, dwell time of 1 s, was performed in 2 min. The sample is tilted by 45°, to free the cutting.

- Filled box was determined with the current of 150 pA in 2 min.

Images of window thinning and U-cut of lamella prepared by FIB is shown in Fig. 3.10.

It is important to notice that in order to achieve a good electron transparency for TEM, the lamellae thickness should be in the range of about 100

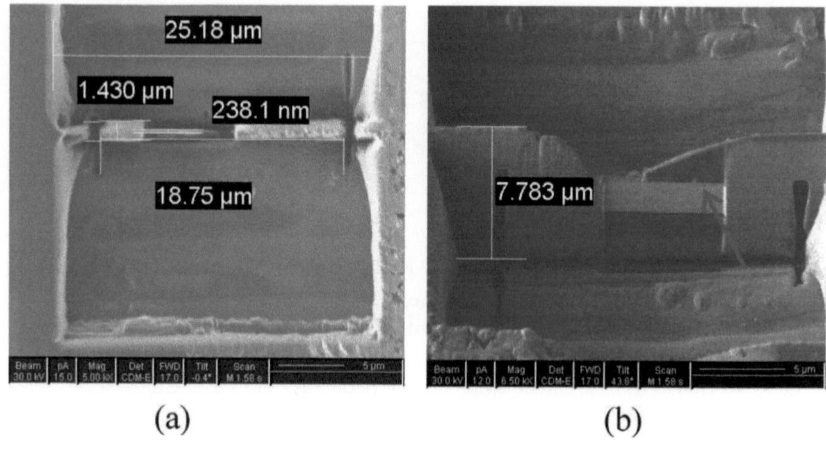

Figure 3.10 Lamella preparation by FIB shows window thinning of the area of interest from (a) top view and (b) side view. U-cuts are shown at both sides of window. A blue arrow points the gate position [Courtesy of U. Zeimer].

nm, which is extremely difficult in the case of the GaN preparation due to the unintentional artifacts introduced by FIB and mentioned above.

After the preparation by FIB or by conventional mechanical grinding and subsequent ion miling the device cross-sections were characterized using a JEOL JEM2200FS transmission electron microscope, operating at 200 kV. In particular, the dislocation analysis was carried out to distinguish between the different dislocation types and to determine the TD densities.

FIB-ing, lamellae transfer and TEM investigations were performed at Humbold University, Berlin [Courtesy of A. Mogilatenko and H. Kirmse].

Chapter 4

Design of experiments

4.1 Motivation

This work focuses on the analyses of the on-set of irreversible degradation effects in GaN HEMTs. A proper analyses of the initial stage of degradations is of particular importance since it is to be expected that the basic physical mechanisms of degradation can be traced back by this method without being disguised by secondary degradation effects. Current drop after several hours operating GaN-based devices often accompanied by the appearance of high gate leakage current are some problems to be clarified. We have established DC-Step-Stress Test and investigated the electrical and physical characterizations before and after stressing. This DC-stress performing at OFF-state has advantage that one can solely see the relation of electrical and mechanical properties and excludes thermal effect (the channel is not open).

FBH produces many wafers with different epitaxy for specific purpose such as more Al concentration in AlGaN layer to have high power device application, AlGaN backbarrier and carbon-doped in GaN buffer to have high break down voltage devices. Regarding these different epitaxy design variations, we selected wafers with systematically varied epi design as tabulated in Table 4.1 to investigate main major issues of degradation.

The wafers were selected to highlight the following dependencies that might affect reliability:

- Mechanical strain in AlGaN barrier layer (Al composition, thickness)

- Influence of GaN cap layer

- Influence of buffer composition and epi growth regime

The results of these experiments design will be shown in chapter 5 according to the following structure:

Table 4.1 Epi design parameters.

Wafer	L_G (μm) gate type	d_{GaNcap} (nm)	$Al_xGa_{1-x}N$ x(%)	d (nm)	d_{buffer} (μm) (Al)GaN	d_{AlN} (nm)	SiC type
"A"	0.5 (T)	-	23	25	2.33 (G)	40	SI
"B"	0.5 (T)	-	23	35	2.31 (G)	40	SI
"C"	0.5 (T)	-	18	25	2.43 (G)	30	SI
"D"	0.5 (T)	5[a]	24	25	2.52 (G)	30	SI
"E"	0.7 (E)	-	23	30	2.4 (G)	360	n
"F"	0.7 (E)	-	23	30	1.84 (Al)[b]	360	n
"G"	0.5 (E)	-	25	25	1.75 (G)	250	n

Remarks:
(T): T-gate, (E): embedded gate
(G): GaN buffer, (Al): AlGaN buffer
SI: semi insulating, n: n-type SiC substrate
a: $7x10^{18}$ Si doped.
b: 5 % Al with GaN channel 15 nm

- Section 1 discussed general findings of study comparison of wafer "A", "B", "C", and "D. Wafer "A" as a reference.

- Section 2 investigated GaN cap influence in deep with optical analysis. Wafer "A" compare to wafer "D".

- Section 3 compared different buffer type: GaN vs. AlGaN buffer on n-type SiC substrate between wafer "E" and "F".

- Section 4 showed investigation on different dislocation densities between wafer "E" and "G".

These studies aim at correlation between epi design parameters and critical voltage V_{CR} as obtained by step stressing measurements. Some wafers with a wide range of V_{CR} are excluded in these investigations.

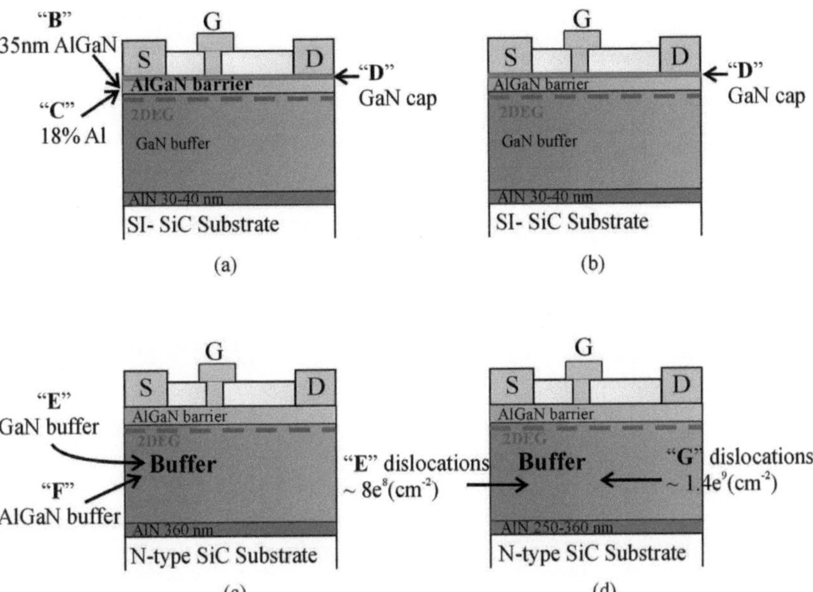

Figure 4.1 Design of experiment to investigate particular areas of interests which are explained in each section in chapter 5 (a) section 1- wafer "A" as a reference, "B" with 35 nm AlGaN layer, "C" with 18 % Al in AlGaN layer, and "D" with GaN cap, (b) section 2- effects on GaN cap (wafer "D"), (c) section 3- buffer comparison, GaN buffer "E" and AlGaN buffer "F" on n-type SiC substrate, and (d) GaN buffer quality between wafer "E" and "D" with higher dislocation densitites.

Chapter 5

Results

This chapter will discuss the results of the investigations described in Table 4.1 in chapter 4. The term of the critical voltage V_{CR} is defined as a threshold voltage of drain-source voltage V_{DS} where one or both, gate leakage and subthreshold drain current increase significantly (> 20 %) and irreversibly during DC stressing at pinched-off conditions. The gate voltage V_G at OFF-state has been set to -7 V to ensure both, complete pinch-off and comparable conditions for all devices. V_{CR} is dependent of epitaxial (epi) designs of GaN HEMTs and determined by DC-Step-Stress tests. Always devices with the same dimension have been stressed: 2x125 μm gate width with source-gate spacing $L_{SG} = 1$ μm, gate-drain spacing $L_{GD} = 6$ μm, gate length $L_G = 0.5$ μm. Variations from this standard are indicated.

The irreversible increases of gate leakage and sub-threshold drain current are considered as degradation. In real device application this might result for example in a reduction of microwave power or efficiency, and can therefore act as an indicator of a starting degradation. It is necessary to investigate epi design concepts towards improved gate leakage and increased device breakdown, and reduced dispersion effects. For this investigation, epi design also includes types of substrates (n-type or semi insulating SiC), nucleation layer AlN (thickness and growth condition), quality and type of buffer (GaN or Al-GaN backbarrier), and Al concentration in AlGaN layer. In this connection, DC-Step-Stress Test act as a fast robustness on wafer tests. Within a short time frame these tests provide an understanding on which epi parameters may play a dominant role.

5.1 GaN HEMTs critical voltage determination by DC-Step-Stress tests

In this section, the comparison of DC-Step-Stress test results of four wafers according to Table 4.1: "A", "B", "C", and "D" are shown accompanied by static IV- and transfer characteristics before and after stressing. Fig. 5.1 represents the evolution of a typical gate leakage and subthreshold drain current during Step-Stress-Test. One can see a trap related charging effect taking place at the initial drain bias steps on devices "A" and "B" (see Fig. 5.1a, 5.1b). It is associated with a recovery of gate leakage I_G and sub-threshold drain current I_D within one step period. However, exceeding a certain threshold voltage, these currents do not recover any more within one step period but starts to increase irreversibly. The onset of irreversible gate leakage degradation and/or subthreshold drain current is marked in Fig. 5.1. We refer to it as the threshold of degradation as V_{CR}. The trap related charging effects are observed again after few steps of point of degradation at higher drain-source voltage V_{DS} (see Fig. 5.2).

Devices of wafer "A" and "B" show an asymmetric (not the same absolute value) increase of gate leakage and subthreshold drain current while devices of wafer "C" and "D" show symmetric increase of both gate leakage and subthreshold drain current until the end of stressing ($V_{DS} = 120$ V). These difference most probably attributes to the punch-through effect in device "A" and "B" where electrons from the source bypass the high electric field area under the gate into the GaN buffer towards the drain contact. This punch-through effect is apparently not present in device "C" with less Al concentration in AlGaN barrier layer and device "D" with GaN cap (see epi design parameters Table 4.1 in chapter 4).

The changes in the electrical performance after step stressing are important indicators for degradation mode interpretation. Fig. 5.3 demonstrates slight changes in IV-output and transfer characteristics after DC-stressing of devices of wafers "A", "B", "C" and "D". A knee-walkout and a slight drop in transfer curves are observed but V_{TH} is unaffected. This knee-walkout is accompanied by an increase of ON-resistance R_{ON} in the low electric field regime. The linear regime is electric field dependent of electron mobility which is affected with phonon and impurity scattering. This means that the knee-walkout and an increase of R_{ON} after stress are associated with an

Main content of this section has been presented in *WOCSDICE*:
P. Ivo, R. Pazirandeh, E. Bahat-Treidel, F. Brunner, O. Hilt, R. Lossy, J. Würfl, G. Tränkle, *GaN HEMTs critical voltage determination by DC-Step-Stress tests*, Workshop on Compound Semiconductor Devices and Integrated Circuits **51** (WOCSDICE), 2008.

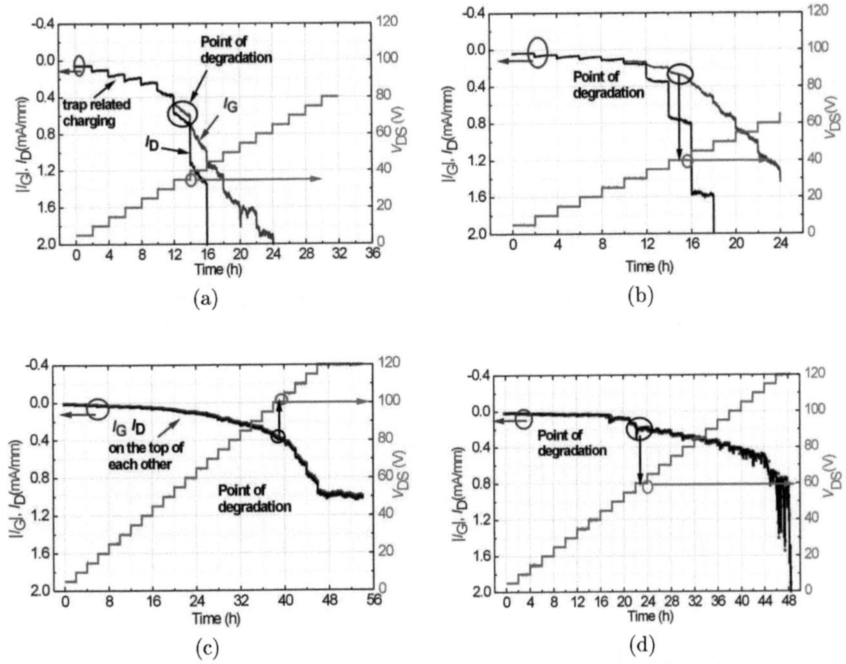

Figure 5.1 Typical DC-Step-Stress Test results of 2x125 μm devices of wafer (a) "A", (b) "B", (c) "C", and (d) "D" with typical $V_{CR} = 20\text{-}35$ V of devices "A" and "B". Devices "C" and "D" have a higher V_{CR} of 40-70 V and 70-100 V respectively. Gate leakage I_G, subthreshold drain current I_D, and drain-source voltage V_{DS} are represented in blue, black and red lines respectively.

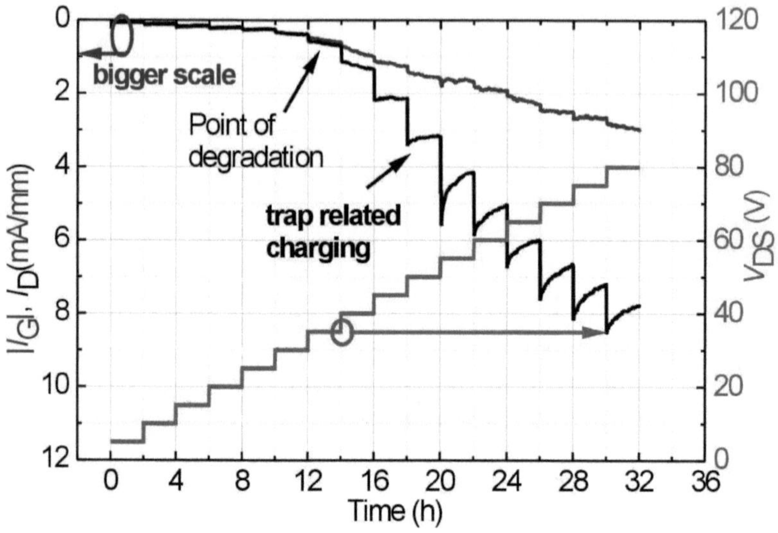

Figure 5.2 Trap related charging effect occurance at higher V_{DS} during DC-Step-Stress tests on device "A". Note: a large scale of I_G and I_D.

increase of this type of scattering. The observed drop of maximum drain current I_{DSS} either indicated a reduction of 2DEG sheet charge density or is due to a decrease of charge mobility. However, these changes already affect the load line for microwave power operation. This is associated with microwave power reduction and therefore degradation of device performance.

It should be noted during initial hours of DC-Step-Stress tests, burn-in possibly takes place. After point of degradation, the irreversible degradation takes place where the gate leakage and subthreshold drain current do not recover to the initial values. The second DC-Step-Stress tests have been performed in the same device. It started with higher values of gate leakage and subthreshold drain current than the first DC-stressing values (see Fig. 5.4). The second DC-stressing also shows more pronounced trap related charging and higher critical voltage when gate leakage and/or subthreshold drain current increase.

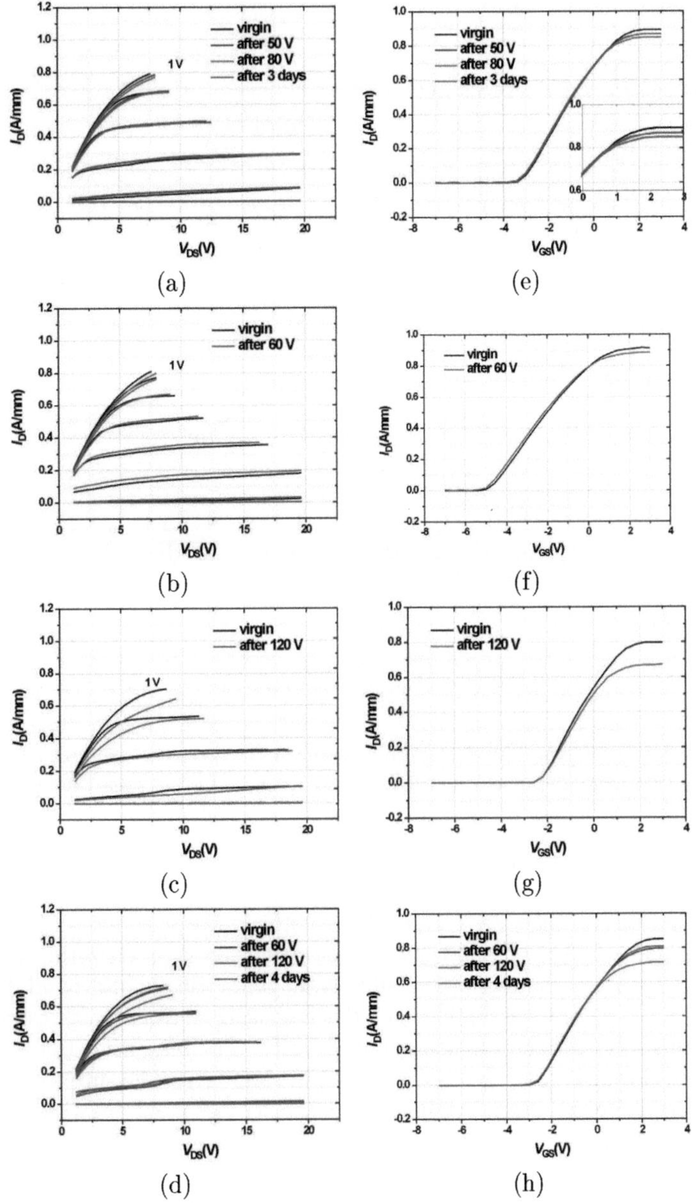

Figure 5.3 IV-output and transfer charateristics of 2x125 μm devices of wafer (a) "A", (b) "B", (c) "C", and "D". The inset in transfer characteristics of device "A" shows that the reduced current after DC-stressing does not recover to its initial value.

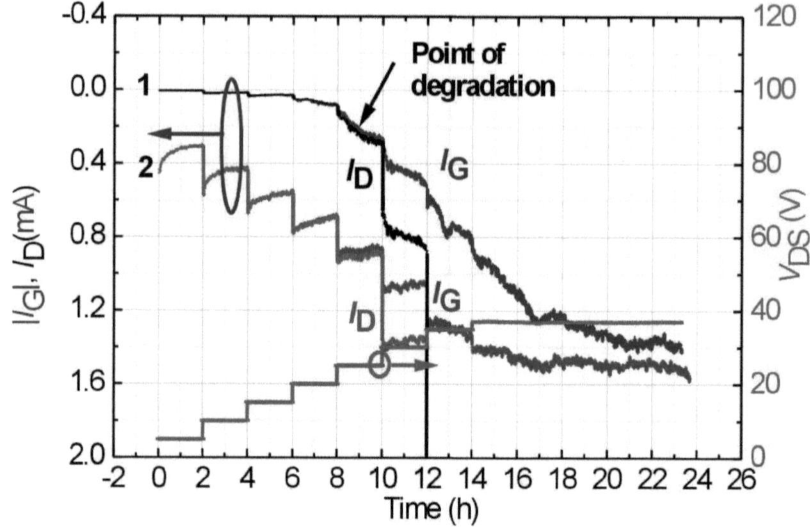

Figure 5.4 First and second DC-stressings marked by 1 and 2 respectively have been performed on device "A". The second DC-stressing starts with higher I_G, I_D values. Their increase per step is quite evident which suggests already on-going degradation effects. The electron punch-through observed during the drain voltage transition from 25 to 30 V is the same for fresh and pre-stressed devices.

5.1.1   Strain in AlGaN layer

For devices of wafer "A" with a barrier thickness of 25 nm and an Al concentration of 23 % and no GaN cap the gate leakage and sub-threshold drain current start to increase irreversibly and symmetrically at around 35 V V_{DS} (see Fig. 5.1a). At 40 V V_{DS}, the gate leakage and subthreshold drain current start to increase non-symmetrically. The reason for this is electron punch-through which will be explained in chapter 6. The subthreshold drain current increases more. At 50 V V_{DS}, the subthreshold drain current partly recovers during one step. This can be interpreted as charging effect, for example the de-trapping of acceptor states. The charge trapping effect within one step becomes more pronounced regarding higher stressing voltage steps.

Wafer "B" with a relatively thicker AlGaN barrier layer of 35 nm as compared to "A" (35 nm vs. 25 nm) is not much different in terms of

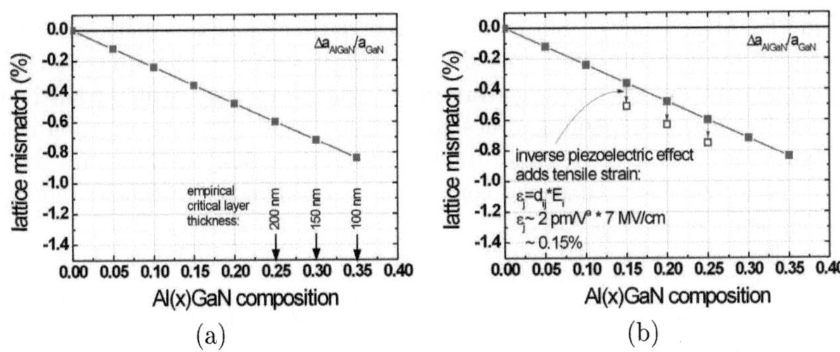

Figure 5.5 Lattice mismatch vs. Al concentration in AlGaN layer with empirical critical layer thickness are marked for each Al concentration (a) and additional tensile strain due to inverse piezoelectriceffect (b).

critical voltage V_{CR} (see Fig. 5.1. b.). The lattice mismatch vs. $Al_xGa_{1-x}N$ composition (see Fig. 5.5. a.) shows that all wafers in Table 4.1 in chapter 4 with Al concentration 18-25% and thickness 15-35 nm are not in the critical thickness range for relaxation of tensile strained AlGaN lattice, therefore one can expect there is no much different results of devices of wafer "A" and "B". However, Ref. [49] simulated 23 % Al concentration in AlGaN barrier has critical thickness ~ 30 nm with dislocation density of 10^{10} cm^{-2}.

In contrast for devices of wafer "C" and "D" having a lower Al concentration in the barrier layer (18 %) and a Si-doped GaN cap (5 nm, 7×10^{19}/cm^3), respectively, gate leakage I_G and subthreshold drain current I_D starts to degrade at higher drain-source voltage V_{DS} of 40-70 V and 70-100 V respectively (see Fig. 5.1. c and d). Both wafers have relatively high break down voltage V_{BR}. It has been found that this is inversely related to the gate leakage current. This has also been discussed in [101].

An interesting recent work showed similar results based on step ramping test of gate-drain voltage V_{GD} [37]. Joh *et al.* pointed out that this kind of degradation may be due to additional tensile strain built-up at the gate edge near the drain contact due to the inverse piezoelectric field. They also found critical voltage values V_{CR} around 20-30 V for their epi designs comparable to the designs "A" and "B". They explained the increasing gate leakage by the onset of hopping conductivity along defects near the drain side gate edge which are created by a local relaxation of the AlGaN barrier layer in the high field regions close to the gate edge.

This model could also explain our finding: devices of wafer "C" have a lower Al (%) barrier concentration of 18 % which gives less mechanical strain

in the AlGaN layer. Since the different lattice constants between AlGaN
and GaN layers create a tensile mechanical strain in the AlGaN barrier,
an increase in Al(%) concentration also increases the tensile strain in the
barrier. Therefore, epitaxial designs with lower Al (%) concentration in the
barrier are less sensitive to additional tensile strain built up due to the inverse
piezoelectric effect [102] and consequently show higher threshold values for
degradation (see Fig 5.5. b.).

Devices of wafer "D" with GaN cap have an increased barrier height [103]
and thus reduced gate leakage current components that are associated with
electron emission or tunnelling across the barrier. Additionally the GaN cap
reduces the critical fields in the vicinity of the gate edge towards the drain
and may therefore postpone potential local relaxation effects. Further study
of device comparison between "epi standard" design ("A") without cap and
with cap ("D") will be discussed in section 5.2 which includes electrical field
simulation under the gate and electroluminescence (EL) measurements.

5.1.2   Gate Technology and Substrate type

The critical voltage V_{CR} measured on devices from wafer "A" with GaN
buffer on SI-SiC and wafer "E" with GaN buffer on n-type SiC are practically the same and are in the range of 30 V. However devices from wafer "E"
show an abrupt increase of both gate and drain leakage as soon as the critical
voltage is reached during step stress testing. On the other hand gate leakage
and subthreshold drain leakage of devices having the n-type substrate is very
low. It is believed that this behavior is due to the fact that the conductive
substrate may act as a bottom field plate which tends to confine electrons in
the channel more efficiently. However, as soon as degradation starts conductive paths form the active region through the buffer might occur.

5.1.3   Summary

We have developed on-wafer Step-Stress-Tests as a fast device robustness
screening method. The method uses irreversible leakage and/or subthreshold drain current increase as an indicator to determine the critical voltage
V_{CR} for device degradation in dependence on epi layer designs for example.
It has been found that epi designs with 23-25 % Al in the barrier layer show
a threshold voltage for degradation of only 20-35 V, whereas the same structures with doped GaN cap or structures with reduced Al concentration (18
%) in the barrier show significantly higher threshold values around 40-70 V
to 70-100 V respectively. These capped-device epi design reduce the high

electric field in the vicinity of the gate at the drain side where the peak of electrical field is present. Less Al concentration devices have less additional tensile strain in AlGaN barrier which also postpones the critical voltage based on criterion. V_{CR} of devices with GaN buffer growth in semi insulating SiC and n-type have no difference in terms of critical voltage V_{CR}.

5.2 Influence of GaN cap on robustness of AlGaN/GaN HEMTs

Following the results of fast screening of different epi design wafers in section 1 it has been found that devices with GaN-cap (wafer "D") have a higher critical voltage V_{CR} than our "standard" epi design without GaN-cap (wafer "A"). Accordingly we investigated further GaN cap influence on robustness more in detail by performing electroluminescence for both devices. In this section, the performance of devices with and without GaN cap has been compared more intensively and correlated to electroluminescence (EL) measurements performed before, during and after DC-Step-Stress testing. Goal of the EL measurements is to provide a possible localization of potentially defective device regions and to gain a better physical understanding of relevant degradation mechanisms [36, 82, 94, 96, 97, 104, 105]. Different physical mechanisms such as thermionic emission, tunnelling, surface or bulk hoping mechanisms and impact ionization hole currents may explain gate leakage and subthreshold drain current increase observed during stress testing.

In order to investigate the influence of a GaN cap, we compared wafers without ("A") and with ("D") an additional GaN cap layer (Si-doped, 5 nm, $7\text{x}10^{19}/\text{cm}^3$). Both structures have a barrier thickness of 25 nm. The average electrical data of both device versions are shown in Table 5.1 where I_{DSS} is the saturated current at a gate voltage of $+1$ V, G_m is the maximum transconductance, V_{TH} is the threshold voltage, and V_{BR} is the breakdown voltage.

Fig. 5.6.a shows a typical gate leakage and subthreshold drain current during Step-Stress-Test for an epitaxial design without GaN cap (device "A"). For device "A", without GaN cap has critical voltage V_{CR} in the range of 20-35 V (see distribution of green circles in Fig. 5.6) . Device

Main content of this section has been presented in *IEEE International Reliability Physics Symposium (IRPS)*:

P. Ivo, A. Glowacki, R. Pazirandeh, E. Bahat-Treidel, R. Lossy, J. Würfl, C. Boit, G. Tränkle, *Influence of GaN cap on robustness of AlGaN/GaN HEMTs*, IEEE International Reliability Physics Symposium, Montreal, 71-75, 2009.

Table 5.1 Selected parameters of wafers.

Device	GaN cap	$I_{DSS} \pm \sigma$ (mA/mm)	$G_m \pm \sigma$ (mS/mm)	$V_{TH} \pm \sigma$ (V)	$V_{BR} \pm \sigma$ (V)	V_{CR} (V)
"A"	No	1073 ±68	240 ± 10	-3.35 ± 0.13	42 ± 3	20-30
"D"	Yes	901 ± 38	233 ± 10	-2.42 ± 0.23	105 ± 9	40-70

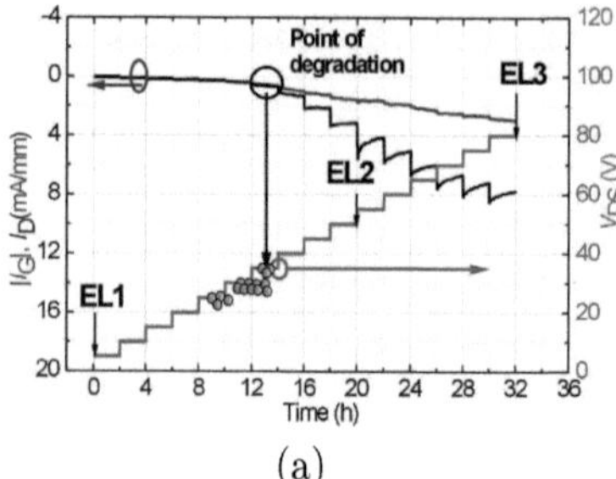

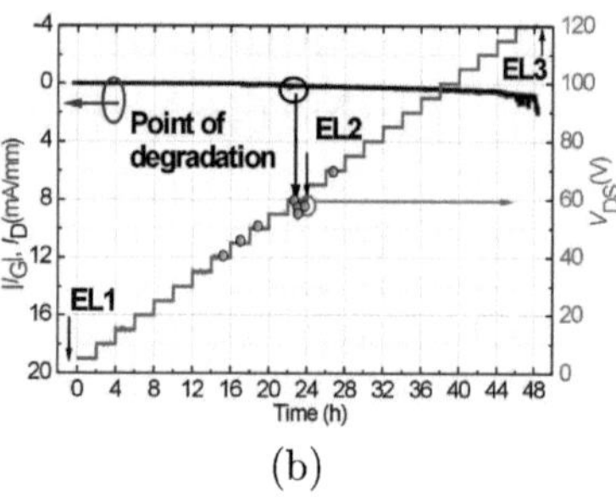

Figure 5.6 A typical DC-Step-Stress test graph of (a) devices "A" (without GaN cap) and (b) devices "D" (with GaN cap) at OFF-state bias conditions (gate voltage V_{GS} = -7 V). The red lines depict the drain voltage V_{DS} over time. The blue and black lines describe the gate leakage current I_G and subthreshold drain current I_D respectively. EL measurements were performed before (EL1), during stressing (EL2) and after stress test (EL3). The green circles represent number of devices that have been stressed with each critical voltage V_{CR} dsitribution.

"D" having a GaN cap has critical voltage V_{CR} in the range of 40-70 V (see distribution of green circles in Fig. 5.6. b).

The sub-threshold drain current of devices "A" starts to increase significantly if the drain voltage exceeds about 30 V. This effect is more pronounced as the drain voltage V_{DS} further increases. We believe that this behaviour is related to the punch-through effect at higher voltage [106]. Any further step-wise increase of the drain voltage provokes a corresponding increase of the sub-threshold current. Within one step the drain current decreases exponentially. This could be an indication of trap charging at these conditions. The higher the drain voltage, the higher the sub-threshold drain leakage current. Electrons that are injected into the GaN buffer can then start to occupy trap levels there and thus reduce the available carrier density in the channel. The occupation of slow traps results in the observed exponential decay of the sub-threshold drain leakage current during one step period.

The IV-characteristics of both devices showed a slight knee walk-out after the first irreversible degradation of the gate leakage has been detected

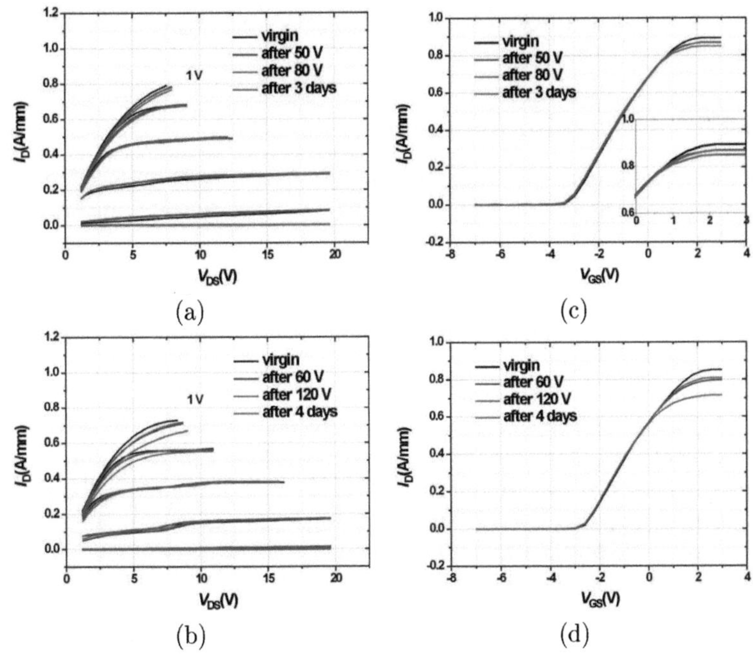

Figure 5.7 The IV- and transfer characteristics of device "A" (a,c) and "B" (b,d) respectively. The black line is before stress, the blue is after stress up to 50 V ("A") and 60 V ("D") , the red is after stress up to 80 V ("A") and 120 V ("D"), and the green is recovery check after several days.

(see Fig. 5.7). More detailed investigations revealed that the knee walk-out and the reduction of I_{DSS} is due to burn-in in the first few hours of device operation. Obviously the irreversible built-up of leakage current does not yet significantly degrade the DC-IV characteristics in addition to device burn-in.

If device step stressing continues up to higher voltages a reversible knee walk-out has been detected. This is visualized especially in Fig. 5.7. b for the capped device which has been stressed up to 120 V. In this case a significant knee walk out was observed initially. However, keeping the device un-biased for a few hours the knee walk-out practically recovers to the value observed after first stressing (see green lines in Fig. 5.7. b). This indicates that high voltage biasing resulted in reversible charging of trap states most probably in the buffer of the device. Similarly, the drain current I_{DSS} for both devices decreases initially after higher voltage stressing and falls back more or less to the first stressed test values during non-biasing for a couple of hours (device

"A" to values of stress up to 50 V and device "B" to values of stress up to 60 V).

5.2.1 Electroluminescence

Measurements of IV-characteristics and EL have been performed before (EL1), during (EL2) and after device stressing (EL3) as indicated in Fig. 5.6. For device "A", the epi design without GaN cap, EL2 and EL3 measurements were performed after stressing up to drain voltage V_{DS} of 50 V and 80 V respectively. The EL2 and EL3 measurements of device "D", epi design with GaN cap, were performed after stressing up to higher drain voltage V_{DS} of 60 V and 120 V since both gate leakage I_G and subthreshold drain current I_D are much smaller than device "A".

EL measurements were carried out before DC-Step-Stressing and after electrical failure detection (irreversible gate leakage and/or subthreshold drain current increase). Measurements taken at OFF-state and ON-state conditions are compared to each other (see Fig. 5.8). EL measurements of non-stressed devices taken at OFF-state always show some bright spots which are more obvious for device "D". Generally at OFF-state conditions the EL images show a more pronounced inhomogeneous behaviour as compared the ON-state conditions. It is estimated that the bright spots result from localized inhomogeneities along the transistor which become dominant at OFF-state conditions. They could be triggered by micro roughness, defect clusters or also process imperfections which might change the local field distribution. Those bright spots become more pronounced during stressing. They show a tendency to coalesce and to increase in number after first and second stress. As indicated in Fig. 5.9 this is accompanied by an increase of the total EL-intensity in the sub-threshold region. It seems that this increase of EL-intensity is roughly correlated to the increasing of gate leakage and/or sub-threshold drain current during stress testing.

The spotty behaviour of EL at OFF-state has been observed as well by Chen et al. [82]. They showed that the bright spots can be related to point defects which may be associated with the existence of donor-acceptor pairs. A dipole configuration of point defect acts as a potential well which can trap some electrons. Zanoni *et al.* pointed out that these bright spots are also representing from current filaments formed after high field stressing. They correlated the appearance of bright spots with the increase of gate leakage current [107]. Those bright spots are interesting to be further investigated by spectral distribution analysis since they obviously appear in device regions where degradation starts initially, extensive material analyses such as TEM

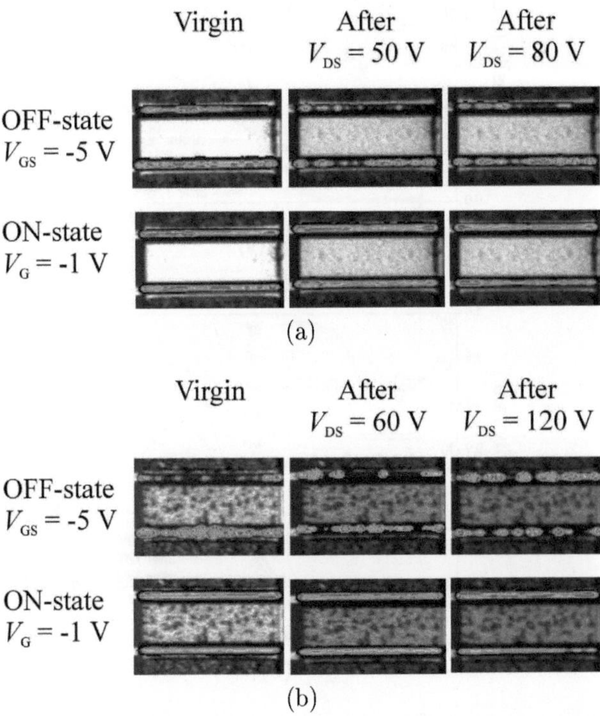

Figure 5.8 Superimposed EL and optical images at $V_{DS} = 10$ V and $V_{GD} = $ -5 V (OFF-state, upper) and $V_{GD} = $ -1 V (ON-state, lower) of 2x125 μm of (a) device "A" - standard epi without GaN cap and (b) device "D" with GaN cap for each point before step-stress test (EL1), during step-stress test (EL2) and after stressing (EL3) as indicated in Fig. 5.6. Note: the images are not yet normalized.

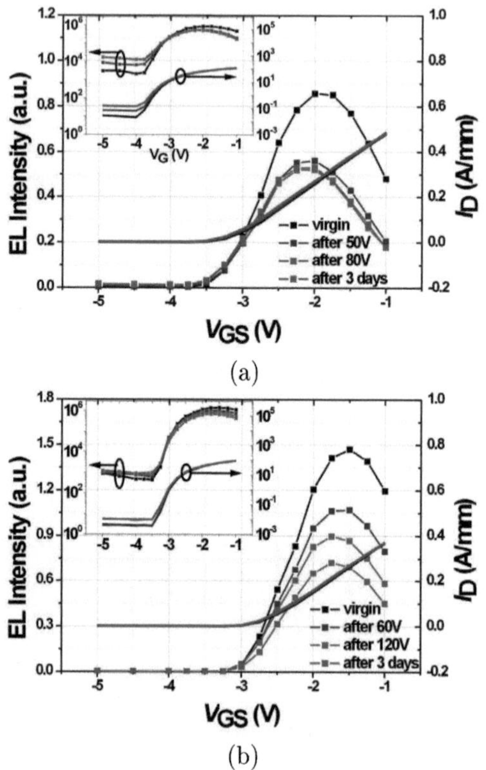

Figure 5.9 EL intensity (left) and drain current I_D (right) versus gate voltage (V_{GS}) of (a) device "A" without GaN cap and (b) device "D" with GaN cap. The black square is before stress, the blue is after stress up to 50 V (for device "A") and 60 V (for device "D"), the red is after stress up to 80 V (for device "A") and 120 V (for device "D"), and the green is recovery check after three days. Insets are the graphs in logarithmic scale which show more pronounced increase of EL intensity and subthreshold drain current after stress at OFF-state. Subthreshold drain current before and after stress up to 50 V (for device "A") and 60 V (for device "D") are lying on the top of each other.

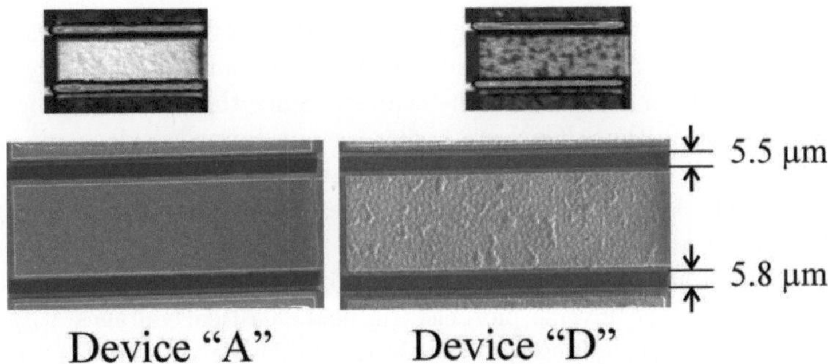

Figure 5.10 Gate-drain spacing of device "A" and device "D". Device "D" has brighter EL intensity (higher field) due to shorter spacing of gate-drain.

cross sections may give further insight on the physical mechanisms associated with these effects.

In contrast to the findings at OFF-state conditions, at ON-state conditions the EL-images show much more homogeneous properties (see Fig. 5.8). The integral intensity of the EL-signal at ON-state conditions during EL measurements always drops for stressed devices and shifts to negative V_{GS} (see Fig. 5.9). After degradation the integral intensity of the EL-signal for devices biased at OFF-state conditions always increases (see insets for both devices "A" and "D" in Fig. 5.9) whereas the opposite is true for devices biased at ON-state conditions during EL measurements.

The asymmetric EL emission between two adjacent fingers as particularly visible for devices "D" (see Fig. 5.9. b) is believed to be due to slight misalignments (200 nm) between the Source/Drain and the Gate electrodes (see Fig. 5.10). Since EL is particularly sensitive to the electrical fields in these device regions any field modifications for example due to geometry changes result in a different EL image.

5.2.2 Electric Field simulation

In order to get some understanding in what might be the reason of the observed effects, simulations of the electrical field distribution in capped and non-capped structures have been performed. They show that capping shields the AlGaN layer in the vicinity of the gate from excessive electrical field (see Fig. 5.11). Since the electrical field in the AlGaN regions close to the GaN cap is more than a factor of two smaller as compared to a similar position in the non-capped AlGaN region, the field stressing of this area is significantly reduced. This finding could support a degradation model based on AlGaN relaxation in the high field regions due the inverse piezoelectric effect as first described by Joh et al. [108]. High values of the electrical field in the AlGaN layer could cause a local relaxation of the AlGaN and consequently provide a leakage path for electrons to the channel. Since the absolute values of the electrical field in the AlGaN are higher for non-capped devices our findings could probably be explained by this mechanism.

Joh et al. [108] indeed showed similar results based on step ramping tests of the gate-drain voltage V_{GD} and pointed out that this kind of degradation may be due to additional tensile strain built-up at the drain side of the gate contact due to the inverse piezoelectric field. Critical voltages V_{CR} of 25-30 V for epitaxial designs comparable to the device without GaN cap had been described. They explained the increasing gate leakage by the onset of hopping conductivity along defects near the drain side gate edge which are created by a local relaxation of the AlGaN barrier layer in the high field regions (~ 7 MV/cm) close to the gate edge. Therefore the magnitude of the electric field in AlGaN layer is a decisive quantity.

5.2.3 Summary

It has been found out that robustness significantly depends on epi layer designs [109]. For example epi design with doped GaN cap shows a higher threshold voltage for degradation of 40-60 V, whereas the same structures without GaN cap shows threshold values of only 20-30V. After degradation the integral intensity of the EL-signal for devices biased at OFF-state conditions always increases whereas the opposite is true for devices biased at ON-state conditions during EL measurements. Generally degraded devices show a larger number of individual EL spots at OFF-state conditions. These spots need to be further investigated for example by cross-sectional FIB and/or TEM analysis in order to explain the appropriate degradation mechanism. The effect of gate length with embeded gate technology and variation of GaN buffer thickness contribute to different values of V_{CR} needs to investigate as well.

Main content of this section has been publisehed in *Microelectronics Reliability*:
P. Ivo, A. Glowacki, E. Bahat-Treidel, R. Lossy, J. Würfl, C. Boit, G. Tränkle, *Comparative study of AlGaN/GaN HEMTs robustness versus buffer design variations by applying Electroluminescence and electrical measurements*, Microelectronics Reliability **51**, 217-223, 2011.

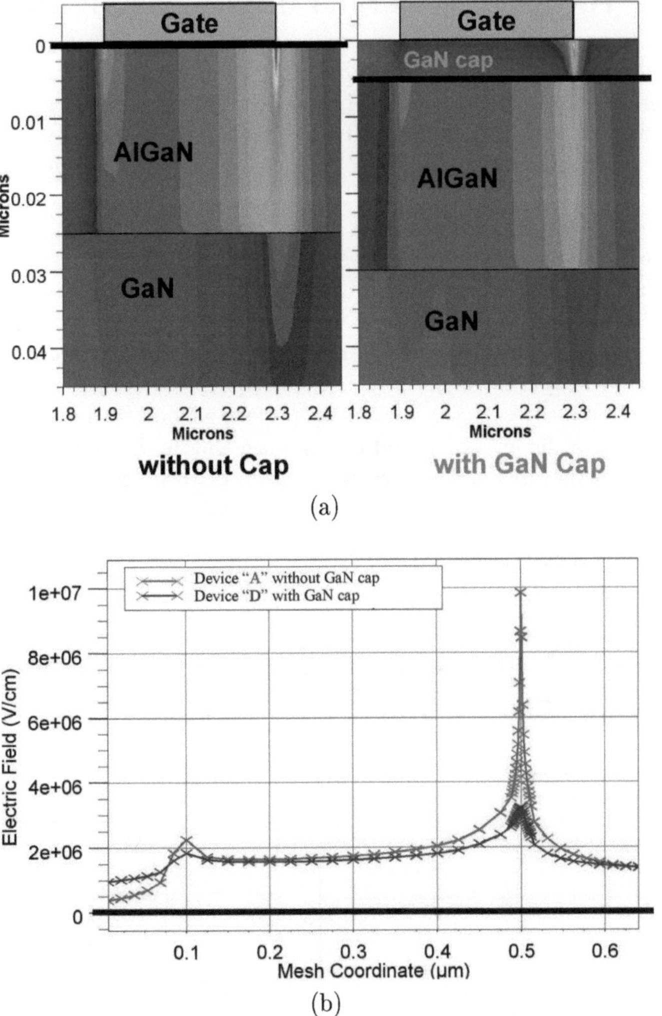

Figure 5.11 High electric fild simulation of devices of wafer "A" without GaN cap and device of wafer "D" with GaN cap (a) cross-section under the gate, and (b) its magnitude along the black lines at (a) [Courtesy of E. Bahat-Treidel].

5.3 Comparative study of AlGaN/GaN HEMTs robustness versus buffer design variations

AlGaN/GaN HEMTs on n-type SiC substrates are of particular interest for power electronic applications. This section compares the degradation properties of devices having different epitaxial buffer designs: wafer "E" with GaN buffer and wafer "F" with AlGaN buffer growth on n-type SiC substrate. Epitaxial designs comprising an AlGaN buffer provide a potential barrier for electrons to the buffer and thus focus electrons in the channel region even at high reverse bias levels (AlGaN back-barrier design). Therefore punch-through breakdown is postponed, breakdown voltage is increased, and sub-threshold leakage current is reduced. These are quite promising properties which make this concept interesting for power microwave and high voltage switching devices.

DC-Step-Stress tests have been performed on two finger devices with 250 μm total finger width and 0.7μm gate length, L_G (see Fig. 5.12). The epitaxial structures of the devices consisted of either a standard GaN buffer (2.4 μm) or an $Al_{0.05}Ga_{0.95}N$ (1.84 μm) back-barrier in combination with a GaN channel layer (15 nm). The top barrier layer consisted of 30 nm thick $Al_{0.23}Ga_{0.77}N$. Both structures had been grown on n-type substrates (see Fig. 5.12). EL has been performed on virgin (EL1), immediately before (EL2) and after degradation (EL3) in OFF-state (V_{GS} = -7 V and V_{DS} = 10 V) for mapping of potentially defective regions and, in dependence on gate bias, to figure out areas of maximum EL-sensitivity.

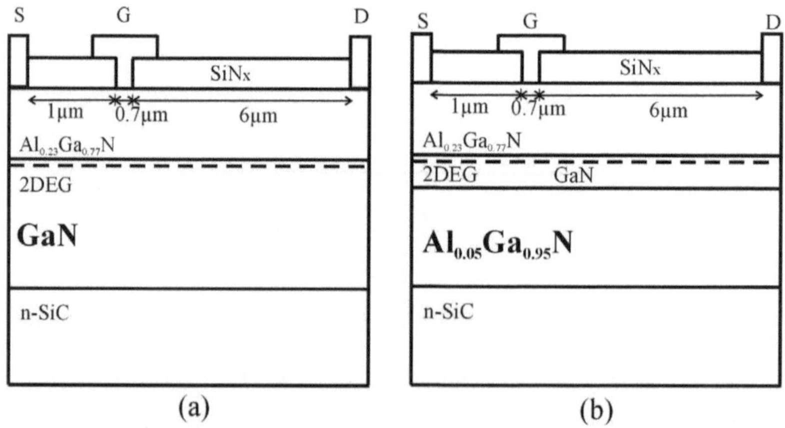

Figure 5.12 A cross-section of AlGaN/GaN HEMTs devices with (a) GaN buffer, wafer "E", and (b) Al0 05Ga0 95N back barrier, wafer "F".

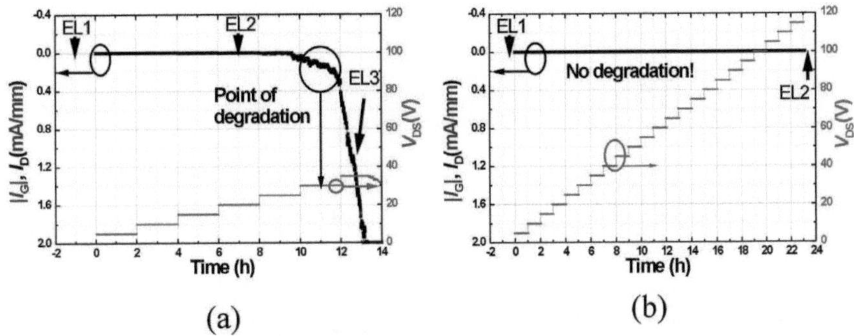

Figure 5.13 A typical DC Step Stress test result of 2x125 μm AlGaN/GaN HEMTs fabricated on: (a) wafer "E" (GaN buffer), and (b) Wafer "F" ($Al_{0.05}Ga_{0.95}N$ back barrier). For both figures gate leakage I_G and subthreshold drain currents I_D are practically identical and are superimposed on top of each other. EL measurements were taken at points marked by EL1, EL2 and EL3.

During step stressing devices "E" with GaN buffer show degradation after reaching a drain voltage of $V_{DS} = 30$ V whereas devices "F" with $Al_{0.05}Ga_{0.95}N$ back barrier show no increase of both I_G and I_D during step stressing up to 120 V (see Fig.5.13). For devices "E" sub-threshold drain and gate leakage increase simultaneously and show exactly the same values. This suggests an electron current flow from the gate directly to the 2DEG region after onset of device degradation (see the degradation of device "E" diode characteristics in Fig. 5.15(a)).

Despite of the observed increase of gate and drain leakage, the output characteristics of devices "E" only show slight knee-walk out (see Fig. 5.14). Devices "F" practically remain unchanged after step stressing tests up to 120 V. The kink effect observed in the knee region of the devices is due to slow traps and only visible if the DC-characteristics is measured at rather small drain voltage increments per time unit (about 1 V/s for the measurements in Fig. 5.14, left side). Pulsed measurements (Fig. 5.14, right side) do not show any kink effect, which means that this effect is not visible in microwave applications.

In general the maximum drain current of wafer "F" with AlGaN buffer structure is reduced by roughly 40% as compared to the same epitaxial design with standard GaN buffer. This is due to the interplay between AlGaN buffer and channel layer thickness which focuses electrons much sharper to the AlGaN barrier/GaN channel interface and therefore reduces the maximum electron concentration in the channel (see chapter 6). The reduction

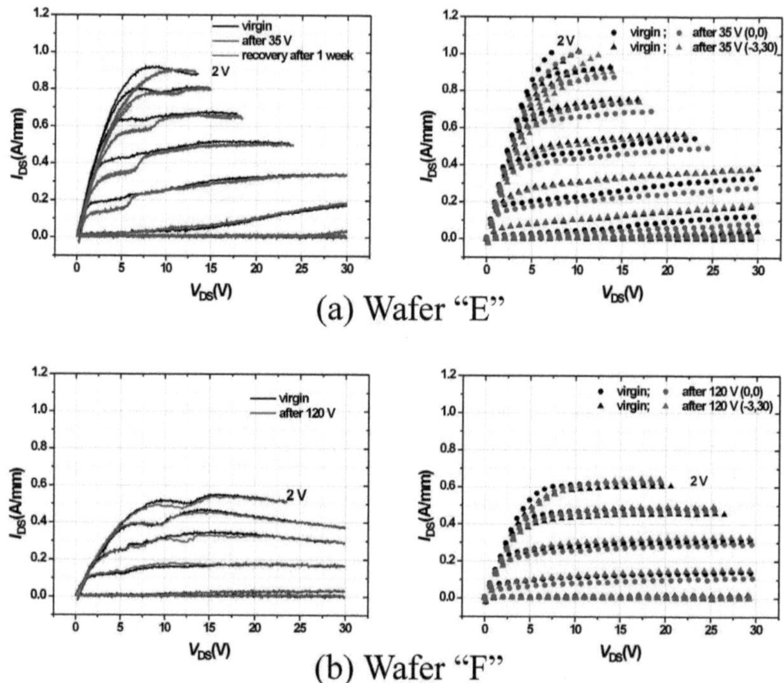

Figure 5.14 Left side: IV static characteristic before marked in (black lines), after (red lines) stressing and recovery check (olive lines) of device "E" (a) and device "F" (b) The output characteristics have been measured at very slow drain voltage increments (1 V/s) to highlight potential kink effects. Right side: Pulsed measurements at certain reference bias points (V_{GS}, V_{DS}) before/after stress: black/red circles at (0,0) and black/red triangles at (-3,30).

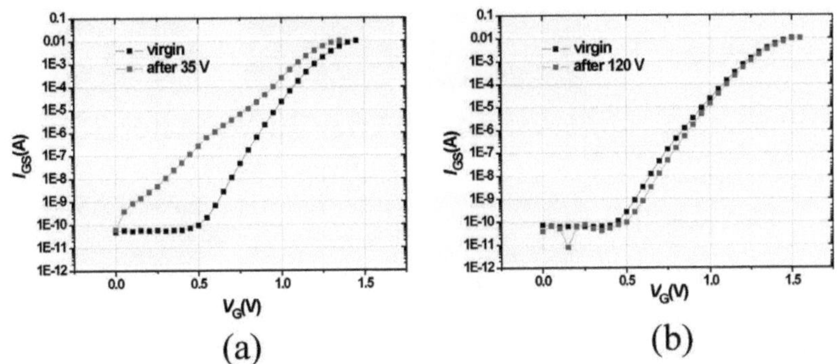

Figure 5.15 Schottky gate IV characteristic before (black squares) and after (red squares) stressing of device "E" (a) and device "F" (b).

of drain current as observed for structures with AlGaN back barrier is not a real disadvantage for implementing this concept. The reduction can be compensated for example by a higher Al concentration in the barrier, by the introduction of an AlN spacer between the AlGaN barrier layer and the GaN channel [110] or by other technological means.

In contrast to the only weak changes in the IV-output characteristics of Fig. 5.14, the Schottky gate IV characteristics in semi log plot, reveals apparent changes that are suggesting permanent degradation effects (no recovery effect to virgin state) associated to either the gate metal itself or the material in direct vicinity of the gate electrode. For devices "F" practically no degradation of the gate diode has been visible, only a slight increase of barrier height could be detected (see Fig. 5.15b). To the contrary, devices form wafer "E" showed a remarkable degradation of the gate diode. The shape of the curve can be interpreted as a voltage dependent resistor in parallel to the Schottky diode. This strongly suggests an additional current path in this region as soon as degradation starts (see Fig. 5.13.a).

5.3.1 Characterization by Electroluminescence

EL measurements at OFF-state conditions support the above-mentioned finding. At virgin devices of wafer "E", few bright spots already appear between the gate and drain region at reverse biased conditions as shown in Fig. 5.16 which are mostly related to native defects i.e. point defects. The bright spots at pinch-off condition are reproducible at constant current and occur at the same places. Chen et al. [82] pointed out that extended defect complexes may be responsible for the spotty behaviour at OFF-state.

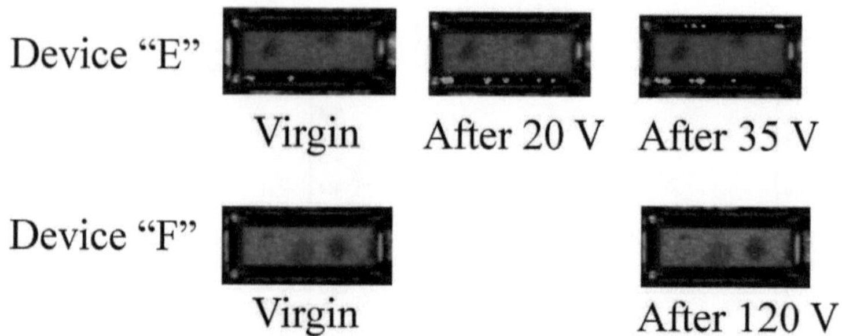

Figure 5.16 Superimposed images show evolution of EL spots at OFF-state (V_{GS} = 7 V and V_{DS} = 10 V) during step stressing. The measurements have been taken as indicated in Fig. 5.13. Note: EL intensity is not normalized.

After stressing, at bias levels of 20 V and 35 V, an increased intensity of the original bright spots are detected and new bright spots occur. We state that this increase of EL intensity is related to defects created after stress due to inverse piezoelectric effect [37] or due to high energetic electron injection in the vicinity of the gate during OFF-state stressing. Both effects may finally open a pathway for electrons to overcome the barrier. Some of these defects are electrically active and result in a modification of the electrical properties of the Schottky diode (see Fig. 5.15). It is believed that as soon such a pathway has formed this may locally reduce the electric field strength, and, if this effect gets pronounced, leads to a drop of EL bell-shaped curve intensity at ON-state conditions at other degraded sample of wafer "E"(see Fig. 5.17a).

At open channel conditions the EL intensity depends on the bias point of the device (see Fig. 5.17). It steeply rises immediately after opening the device from pinch off conditions and reaches maximum values for semi open conditions at roughly $1/4\,I_{DSSmax}$. Since the electrical fields in the gate region reach a maximum value close to pinch off, this behaviour suggests strong field dependent light generation mechanisms at these conditions such as intra band electron transition [98]. When gradually turning on the device from pinch-off, the electron density increases while the fields are still quite strong. This gives rise to a strong increase of the EL signal. If the device is biased further into open conditions the E field in the channel decreases and thus the probability for intra-band transitions also decreases despite the fact that more and more electrons are taking part in the transport mechanism. This means that the EL intensity at open bias conditions close to pinch-off

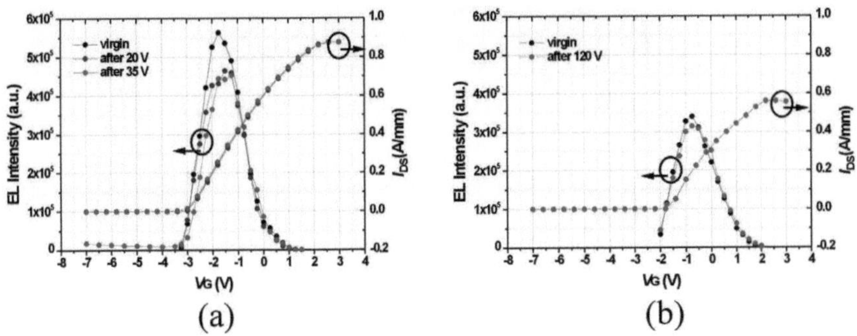

Figure 5.17 EL peak distribution on transfer characteristics at $V_{DS} = 10$ V of device "E" (a) and device "F" (b).

is very sensitive to the magnitude of the electric field in the gate region. Therefore it acts as a powerful indicator of electric field changes close to the gate region during degradation. As pointed out above, degradation of sample "E" is accompanied by the appearance of a parallel resistive part to the gate Schottky diode. This may be interpreted as an accumulation of defects (potentially point defects according to [82]) which in-turn locally reduce or modify the electric field and thus lead to a reduction of the over all EL intensity in open conditions close to pinch off as shown in Fig. 5.17. Additionally the "turn-on slope" of the EL signal is shifted to more positive values of V_{GS}. This means that due to a reduction of the electric field close to the drain side edge of gate a larger electron concentration in the 2DEG is now needed to create the same EL signal as compared to the unstressed case. For device "F" only a slight reduction of the EL signal has been detected after step stressing (up to 120 V in this case).

Fig. 5.18 shows the evolution of EL patterns of a degraded sample in dependence on gate voltage at a constant drain bias of 10 V. Note that the EL intensity is not normalized which means that bright spots of the same colour from adjacent images do not necessarily have the same intensity. It is only possible to directly compare the EL spectra taken at $V_{GS} = -7$ V and $V_{GS} = 1$ V since at these values the normalized spectra according to Fig. 5.17 roughly show the same values. At a gate bias of $V_{GS} = 1$ V only the bottom finger emits light whereas the top finger is already dimmed. Furthermore it can be seen that the right quarter of the top finger is completely blanked out exactly at those positions where bright spots appear at OFF-state conditions. This indicates that at these positions an extensive defect creation might have taken place which may have locally reduced the electric field and thus

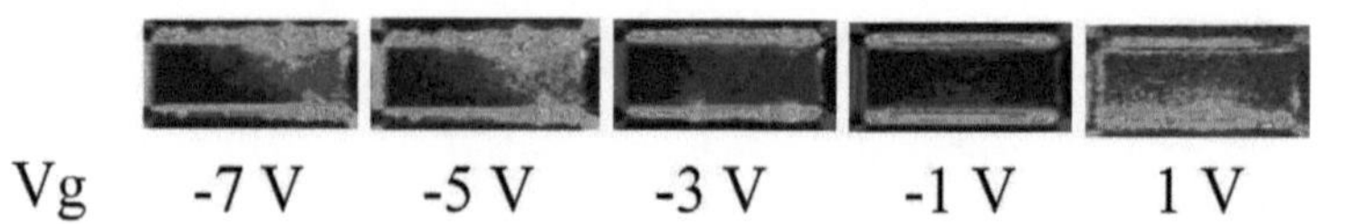

Figure 5.18 EL spectrum of degraded device of wafer "E" (as normalized EL intensity is shown in Fig. 5.17) after 35 V stress showing the evolution of EL intensity at different gate voltages, V_{GS}. The drain voltage V_{DS} was kept constant at 10 V during these measurements. Dark colours i.e. blue indicates lower intensity/electric field and bright colours i.e. red characterizes regions of higher intensity/electric field.

inhibited EL in these regions. Since this correlation between bright spots at pinch-off conditions and dark spot at ON-state does not apply for all bright spots observed at pinch-off we believe that the observed effect is not a general rule and that it represents a very specific mode of degradation. Further investigations on other samples have shown that the observed effect is much more pronounce if a very strong device degradation is enforced during DC-Step-Stress testing.

It can be seen that the emission from the top finger is lagging behind the bottom finger at ON-state condition. This may be due to a processing issue, most probably to a slight misalignment of the gate fingers in the two transistor sections. At $V_{GS} = - 3$ V (at threshold voltage) the situation is reversed. Most of the area of the top finger already emits light which means that at this particular bias voltage the top finger is already biased slightly in on state whereas the bottom finger is still pinched off.

5.3.2   Simulations of band diagrams and internal electric field distribution

Two dimensional physical device simulations (Silvaco-Atlas) were performed [Courtesy of E. Bahat-Treidel] to calculate the band diagram of devices from wafers "E" and "F" (see Fig. 5.19a). It is clearly seen that that device "F" (with back barrier, indicated as red lines) provides an efficient potential barrier for electrons to prevent them from being injected into the buffer. On the other hand the steep slope of the conduction band results in a reduction of electron population in the channel [106] which gives rise for the reduced maximum saturation current as observed for back barrier wafers if all other conditions had been kept constant.

The internal electrical fields at OFF-state conditions for devices from wafer "F" are significantly smaller as compared to devices from wafer "E"

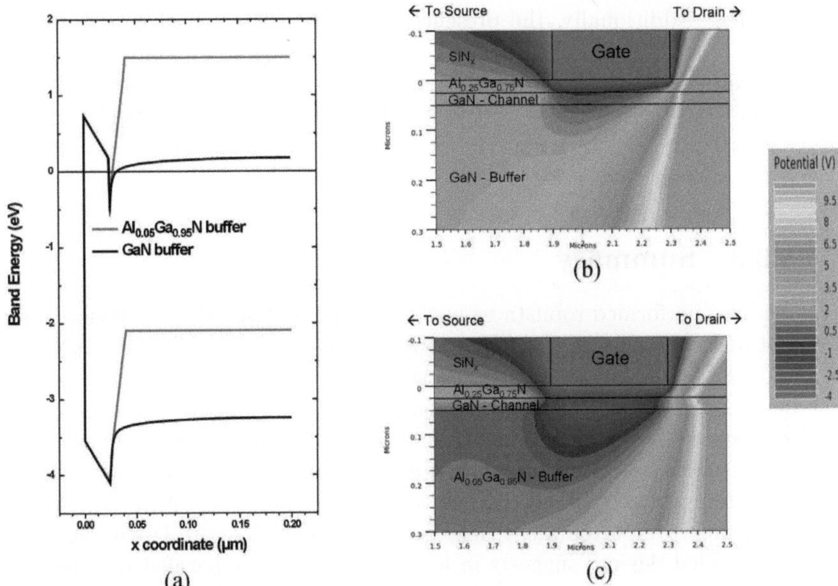

Figure 5.19 (a) Band diagram simulation of device from wafer "E" with GaN buffer (black lines) and device from wafer "F" with AlGaN buffer (red lines). Wafer "F" has a steep slope of conduction band which confines sheet electron concentration in 2DEG channel. Comparative simulation of potential distribution in the vicinity of the gate electrode at pinched-off conditions (V_{GS} = - 6 V, V_{DS} = 45 V): (b) Devices from wafer "E" and (c) wafer "F" with AlGaN buffer and 30 nm GaN channel in this case. The electric field of device from wafer "F" is roughly half of device from wafer "E" [Courtesy of E. Bahat-Treidel].

if calculated at the same biasing conditions (see Fig. 5.19b and c). This is mainly due to the modified band structure in presence of an AlGaN back barrier and, in open conditions, also due the lower electron density in the channel (compare max. drain current values in Fig. 5.14) [106]. Since we have strong indications that the observed degradation of device "E" occurs in the high field region of the gate, the better device robustness for devices from wafer "F" may in part be attributed to local E field reduction in the gate vicinity. Additionally, the presence of a bulk AlGaN back barrier reduces the tensile strain in the AlGaN barrier layer of devices from wafer "F" as compared to devices from wafer "E" and thus shifts the relaxation limit of the AlGaN barrier to higher Al values. This means that higher fields would be necessary to provoke a material degradation due to the inverse piezoelectric effect as suggested in [37].

5.3.3   Summary

We have performed robustness measurements on GaN HEMT devices manufactured on wafers with different buffer designs: wafer "E" with GaN buffer and "F" with AlGaN buffer. A superior stability of AlGaN back-barrier devices on n-type SiC substrate has been demonstrated. After step stressing devices from wafers with GaN buffer depict an abrupt gate leakage and subthreshold drain current increase after stepping up to $V_{DS} = 30$ V while devices from wafer "F" shows no gate leakage and/or subthreshold drain current practically even after 120 V stressing. The Schottky gate IV-characteristics of degraded devices suggests in a voltage dependent parallel resistor which is interpreted as an additional current path through the barrier as a result of degradation. EL measurements show an evolution of bright spots for devices from wafer "E" which possibly relate to the defects created during the stress. Degraded devices show a lower integral EL intensity as compared to the virgin state. This indicates that defect creation results in a reduction of the electric field in the affected regions which therefore leads to a reduced EL intensity since the threshold for the intra-band emission may not be reached anymore or only at higher gate or drain bias levels. Partly, defects occurring at OFF-state condition lead to dark spots in ON-state condition which also calls for a reduction of electric field in these particular regions after degradation.

5.4 Degradation mechanisms of GaN HEMTs in dependence on buffer quality and gate technology

AlGaN/GaN HEMTs on n-type SiC substrates are of particular interest for power electronic applications. Therefore different buffer structures have been grown and compared to each other with respect to the degradation modes of devices fabricated on these materials. Devices fabricated on thin and thick buffer layers ("E" vs. "G") structures were stressed. The analysis was accompanied by electrical measurement of DC and pulsed device characteristics and complemented by electroluminescence (EL) characterization to localize defective device regions before and after degradation tests. AlGaN/GaN HEMTs consisting of 2x125 μm-devices with 6 μm gate-drain distance have been fabricated on MOVPE grown wafers with different GaN buffer layer thicknesses of 2.4 μm and 1.7 μm GaN respectively on n-type SiC substrate (for comparison see Table 5.2). The epi curvature during GaN deposition of wafer "E" and "G" are shown in Fig. 5.20. Wafer "E" and "G" has different temperature growth of AlN deposition (marked with dashed yellow lines in Fig. 5.20). Brunner *et al.* showed that high temperature AlN growth provides a low number of dislocation density in GaN buffer layer [48]. Since wafer "G" has low temperature AlN growth, it ended up with relatively high edge dislocation density $\sim 1.4\mathrm{x}10^9$ cm^{-2}. This is also possibly due to relatively thin GaN buffer thickness of wafer "G" (1.7 μm).

Fig. 5.21 compares typical DC-Step-Stress test results obtained from devices with different buffer layer thicknesses and qualities. Devices from wafer "E" show an increase of both gate leakage and sub-threshold drain current simultaneously at a critical voltage of 30 V (see Fig. 5.21a). In contrast, devices from wafer "G" show a different way of degradation: the sub-threshold drain current increases abruptly at a higher critical voltage of $V_{DS} = 55$ V whereas the gate leakage current is not affected. It is about three orders of magnitude lower as compared to devices from wafer "E" at high drain voltage (see Fig. 5.21b). Degradation is accompanied by a significant reduction of drain current after stressing (see Fig. 5.22b). As mentioned in the last section, the gate Schottky diode measurement indicates a parallel resistor occurrence after stressing on devices of wafer "E" (see Fig. 5.22c)

Main content of this section has been presented in *ROCS 2011*:
P. Ivo, Ute Zeimer, P. Kotara, E. Cho, L. Schellhase, E. Bahat-Treidel, J. Würfl, and G. Tränkle, A. Glowacki, and C. Boit, *Degradation mechanisms of GaN HEMTs in dependence on buffer quality and gate technology*, Reliability on Compound Semiconductor (ROCS), 2011.

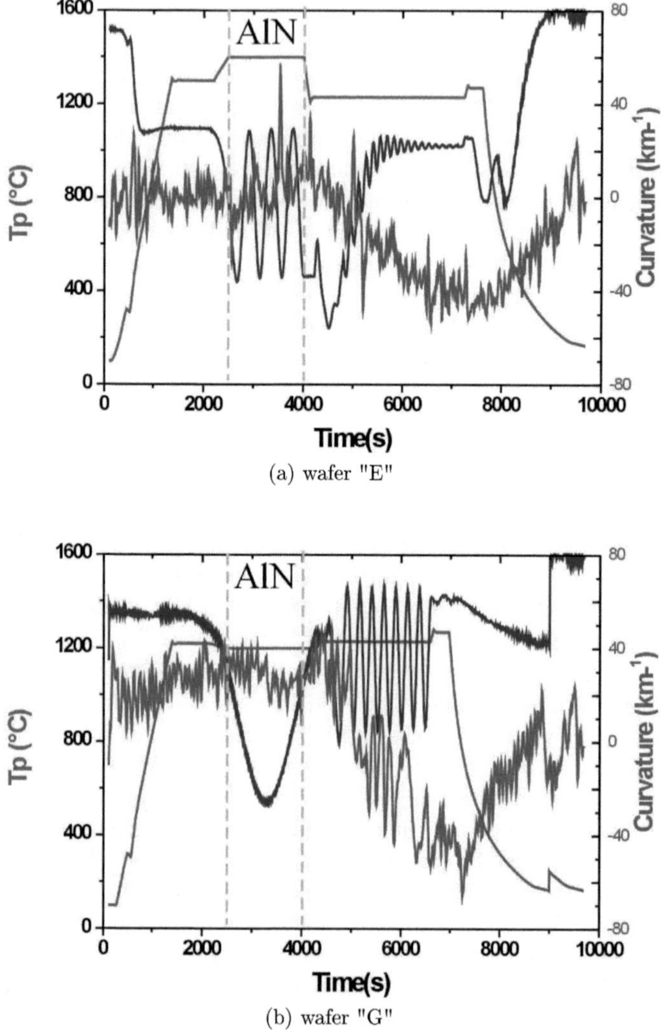

(a) wafer "E"

(b) wafer "G"

Figure 5.20 Epi curvature of GaN growth of (a) wafer "E", and (b) wafer "G". The temperature AlN growth of wafer "E" is ~ 1400 °C and wafer "G" is ~ 1200 °C.

Table 5.2 Epi-designs under investigation.

Wafer	d_{GaN} (μm)	d_{AlN} (nm), T ($^\circ$C)	$d_{barrier}$ (nm)	L_G (μm)	Defect densities (cm^{-2})	
					edge	screw
"E"	2.4	360, HT*	30, Al$_{0.23}$Ga$_{0.77}$N	0.7	~ 8x10^8	~ 7x10^7
"G"	1.7	250, LT*	25, Al$_{0.25}$Ga$_{0.75}$N	0.5	~ 1.4x10^9	~ 7x10^7

Remarks:
*HT/LT: high/low temperature growth

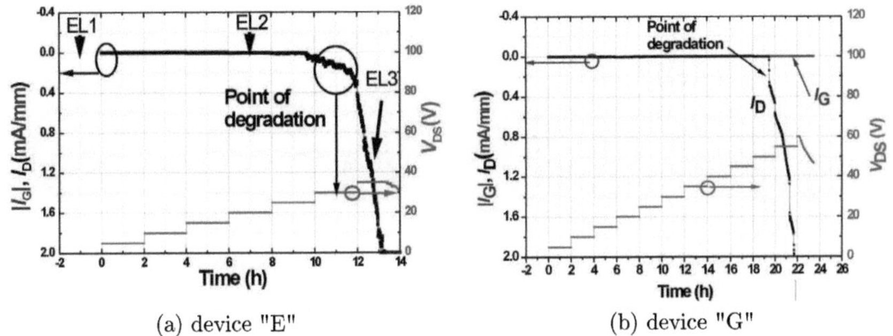

Figure 5.21 A typical drain ramping DC Step Stress test result of 2x125 μm Al-GaN/GaN HEMTs at pinch-off V_{GS} = -7 V (a) device from wafer "E" [111] and (b) device from wafer "G". In device from wafer "E" gate leakage (blue line) and drain current (black line) are lying on top of each other.

while a reduction of barrier height from 0.9 to 0.7 eV occurs on devices of wafer "G". However, the log-linear region of the forward biased Schottky diode is not affected (see Fig. 5.22d). Devices of wafer "E" tend to start degrading at the gate diode first: the electrical characteristics of the gate diodes severely degrade resulting in a direct leakage current flow from the gate to the drain via the 2DEG region at the gate (see also Ref. [111]). This results a strongly correlated simultaneous increase of gate and drain current during step stressing tests (see Fig. 5.21a).

For drain voltage ramping at pinched-off conditions, devices from "G" degrade by creating a current path, most probably in the buffer region, that bypasses the high field gate area underneath the channel. There are indications that a defect assisted electron transport takes place which may

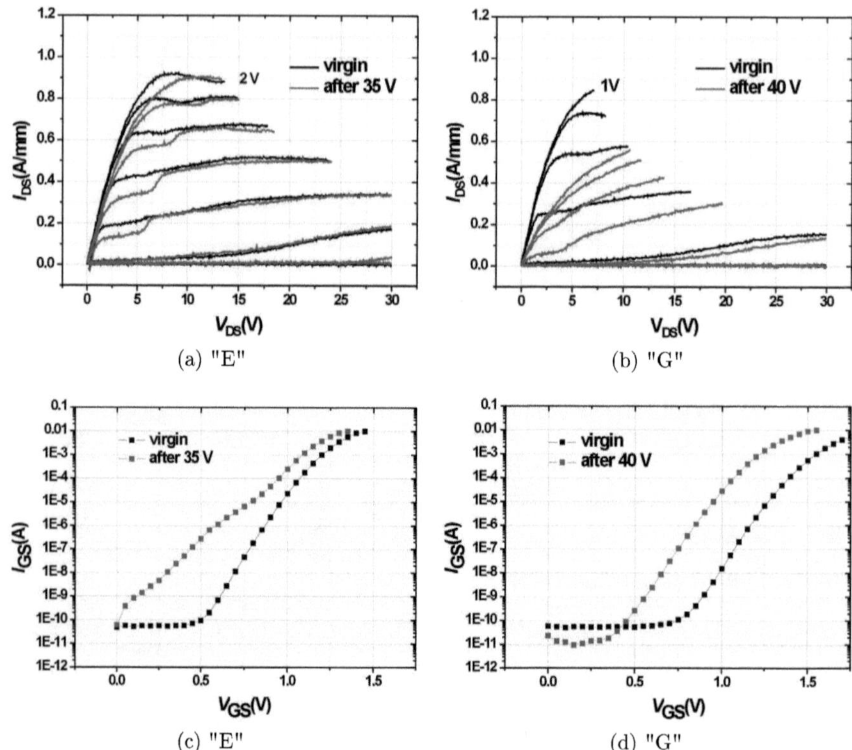

Figure 5.22 IV- and diode characteristics before and after stressing (a) (c) device from wafer "E" [111] and (b) (d) device from wafer "G", respectively.

incorporate fixed negative charges close to the channel. As a consequence, the maximum drain current is then permanently reduced. At the same time the barrier height also changes and reduces, although the gate leakage remains small. This could be due to charge build up underneath the gate in the AlGaN barrier or in the buffer as a consequence of degradation leading to charge redistribution immediately under the gate electrode (Fermi level pinning).

5.4.1 Electroluminescence

Depending on buffer construction the EL-images of degraded samples show different fingerprints (see Fig. 5.23). Normalized emission images of the

EL signal of devices from wafer "G" show a significant EL intensity even if the devices are driven in complete ON-state operation conditions whereas degraded samples on devices from wafer "E" completely dim out at ON-state conditions.

At virgin state the normalized emission images of the EL signal show a bell-shaped curved over transfer characteristics. This has also been observed by other groups and is attributed to intraband transition in the 2DEG [98] (see black lines in Fig. 5.23a and 5.26b). The luminescence by definition is the product of nonequilibrium, and has a radiative rate R = n_u n_1 P_{ul} where n_u and n_1 are the densities of carriers in the upper and lower states respectively, and P_{ul} is the probability for one carrier/cm^3 in the upper state to make a radiative transition to one vacancy/cm^3 in the lower state [93]. The abrupt increase of EL intensity immediately after opening of the channel is due to the simultaneous presence of carriers and empty lower states to which radiative transitions are likely. As the channel further opens the number of carriers increases and the electrical field in gate vicinity reduces. Therefore the probability for radiative transitions decreases. However, if additional energetic states are created, for example by defect creation during degradation, the probability of radiative transitions may still be quite high- thus the EL intensity may not dim out at high channel drive levels for degraded devices. This effect is seen for devices from wafer "G" after degradation. Then the EL intensity peak drops and shifts to positive gate voltage but does not dim out (see Fig. 5.23d). Carrier flow is now strongly determined by traps created during stressing which in turn, due to the charge balance in channel vicinity, causes a significant reduction of maximum drain current as seen in Fig. 5.22b.

The locally resolved EL images in dependence on gate bias according to Fig. 5.23b and 5.23d show comparable images for the virgin devices. However, after degradation a completely different behavior can be stated. After stressing, the EL images of devices from wafer "E" at OFF-state (V_{GS} = -7 V) show a significant increase in number and size of bright spots whereas in the particular devices from wafer "G" only a single bright spot appeared. In contrast, the EL images of the wafer "E" type devices dim out at ON-state (V_{GS} = 2 V), while the device from wafer "G" showed numerous bright spots.

In order to get further insight into degradation mechanisms, FIB cross-sections of the degraded and non-degraded devices have been performed. FIB trenches have been cut through nominal bright (E1) and dark (E2) EL spots of degraded devices "E" (see Fig. 5.24). However, no real differences could be observed after SEM inspection on non-degraded and degraded devices of wafer "E". As visible in Fig. 5.24 there is no particular difference in FIB

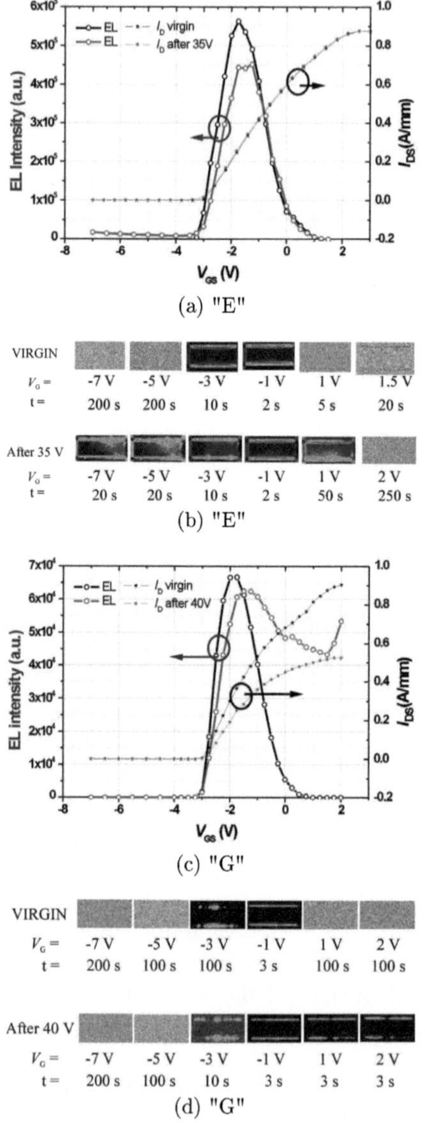

(a) "E"

(b) "E"

(c) "G"

(d) "G"

Figure 5.23 Normalized EL intensity at $V_{DS} = 10$ V over transfer characteristics at virgin state (black lines) and after stress (red lines) of (a) device from wafer "E" [111] and (b) device from wafer "G". Locally resolved EL evolution in a two-finger test device at $V_{DS} = 10$ V in dependence on gate voltage (c) device from wafer "E" [111] and (d) device from wafer "G"

cross-sections. This means that the defects that are created are probably beyond the detection limits of SEM. However, two distinct features can be noticed after the FIB preparation of both non-degraded and degraded devices: an extremely rough gate surface and small voids at the both sides of gate walls. In order to prove if these features are caused by the FIB preparation or they are initially present in the fabricated devices, we carried out an additional conventional TEM specimen preparation of the non-degraded device of device from wafer "E". Scanning transmission electron microscopy (STEM) image in Fig. 5.25 shows the T-gate contact conventionally prepared by mechanical grinding followed by subsequent ion milling. The image was obtained using the so-called high-angle annular dark-field (HAADF) detector. The intensity distribution in the STEM HAADF image strongly depends on the average atomic number of the analysed material as well as the specimen thickness. Thus, the possible rough specimen surfaces can be clearly visualized using this technique. As visible, the conventionally prepared gate in Fig. 5.25 shows a smooth cross-section indicating that the surface roughening effect is an artifact of the FIB preparation. In contrast, the small voids appearing at the sides of the walls are observed for both cross-sectioning methods, indicating that they are initially formed during the device fabrication.

Since SEM analysis did not reveal any differences in the structure of degraded and non-degraded devices, further TEM investigations were performed on non-degraded devices of wafer "E". Fig. 5.26 shows two annular dark-field (ADF) STEM images of non-degraded and degraded devices. In contrast to the Z-contrast appearing in HAADF STEM, the ADF images exhibit a strain contrast. Consequently, it can be used to visualize threading dislocations in the layers. It is impossible to distinguish between the screw and edge dislocations using these images. However, this imaging technique gives precise information on the total dislocation density in the devices. According to the STEM analysis, still we observed no significant difference between bright and dark areas at OFF-state during EL measurements (see Fig. 5.26). However, this result is contradictory to other group observation. Ref. [81] observed higher dislocation density at the bright spots of EL measurements at OFF-state. According to the results of our analysis, dislocation are randomly distributed over the GaN buffer. Thus, we are not sure to suspect dislocation density as the main cause of degradation after stress. Obviously, a thinner GaN buffer with a thickness of less than 1.7 μm has a higher defect density as indicated in Fig. 5.26a by the dashed line.

However we found, that both degraded and non-degraded devices from wafer "G" showed extended voids at the gate foot sidewalls (see Fig. 5.27). This is attributed to a processing problem. Since the gate metal is not touch-

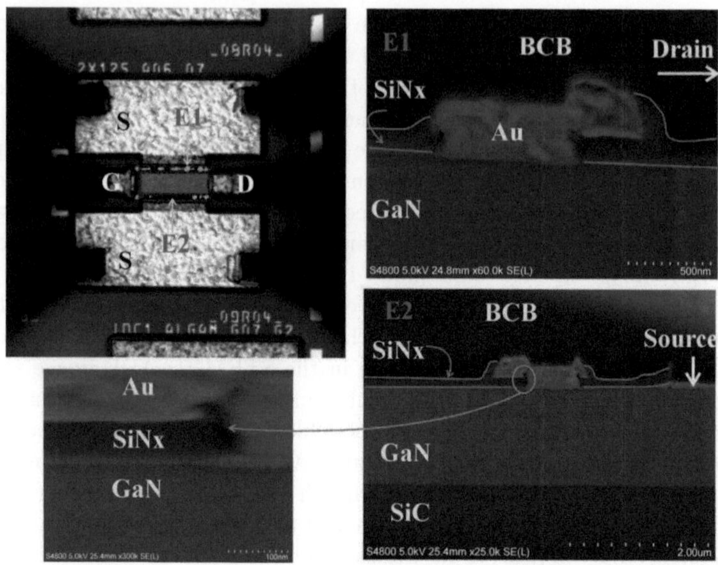

Figure 5.24 FIB cross-section of degraded device "E" at bright (E1) and dark (E2) areas at OFF-state luminescence which show voids at the gate wall for both areas. The wings of gate are assymmetric and some artefacts due to ion beam during FIB-ing are present [Courtesy of P. Kotara].

ing the SiN_x passivation any more (small gap), this region is not passivated anymore. Thus electrons can charge up the area around the gate leading to a virtual field plate and thus reduce electric field intensity at the drain side edge of the gate. This would explain the lower leakage compared to devices from wafer "E".

As shown in Fig. 5.21, the devices from wafer "G" show a significantly different mode of degradation as compared to devices from wafer "E". They do not degrade by creating a gate leakage path, instead a drain current path appears, that cannot be properly controlled by the gate anymore (similar to punch through effect). Therefore, we believe that the dominant part of devices from wafer "G" degradation is finally due to the presence of a comparably high density of vertical defects in the epi stack in connection with GaN buffer quality [48] (see defect densities in Table 5.2).

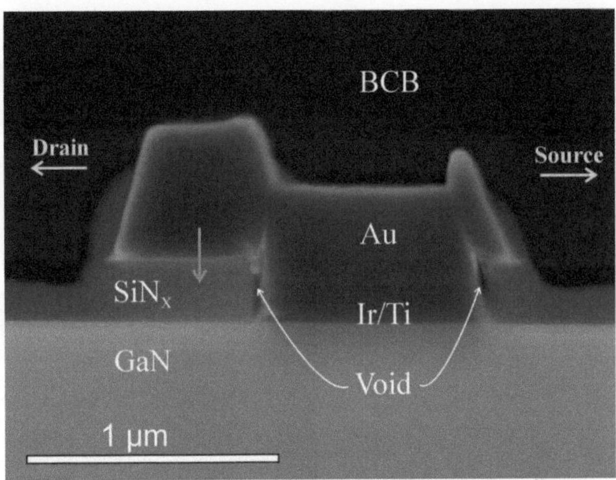

Figure 5.25 Mechanical grinding cross-section of non-degraded device "E". Voids are also present at both gate walls with mechanical grinding cross-section [Courtesy of A. Mogilatenko].

5.4.2 Summary

It is apparent that devices of AlGaN/GaN HEMTs fabricated on thicker buffer structures with less edge dislocations ($\sim 8x10^8$ cm^{-2}) tend to degrade in close vicinity of the gate by opening a gate leakage path through the Al-GaN barrier [111]. In contrast, devices on thin buffer layer material show a drain current increase during step stressing whereas gate leakage is not affected. This mode of degradation is attributed to electrons by-passing the gate region via defect-assisted conductivity through the buffer layer underneath the gate and the drain side channel region. At the same time the total available device current drops significantly and large EL signals can be seen even at fully opened channel conditions. The appearance of the EL signal at ON-state conditions together with a significantly reduced drain current indicates that permanent trapping states are created during degradation. In conclusion it has been shown that device degradation modes significantly depend on buffer quality of epitaxial design and gate technology. Wafers with high GaN buffer defect densities may provoke defect related charge trapping effects after stressing, leading to strongly reduced drain currents of degraded device due to punch-through. On the other hand it seems that pure gate

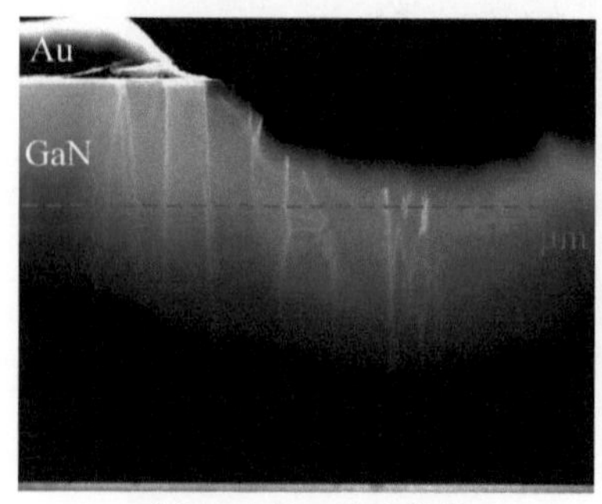

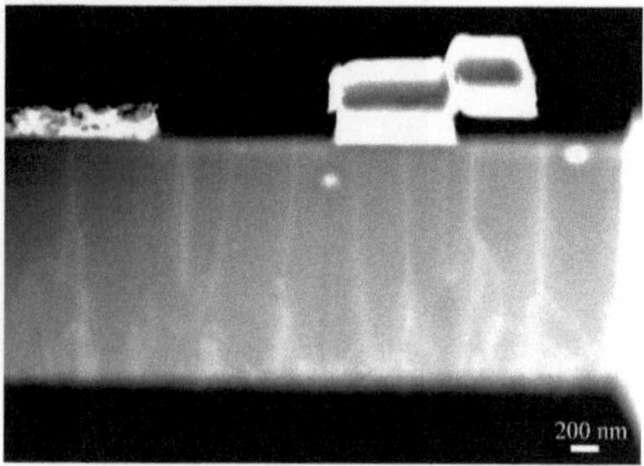

<table>
<tr><td>(a) non-degraded "E"</td><td>(b) dark spot degraded "E"</td></tr>
</table>

Figure 5.26 TEM images of area under the gate of devices "E" of (a) non-degraded device and degraded device at (b) dark area. Notice that two bright circles are artifacts from FIB-ing [Courtesy of A. Mogilatenko].

assisted degradation effects are postponed if the gate leakage is reduced due to unpassivated surface. Apparently, the "void" due to process might be "useful" to block the gate leakage.

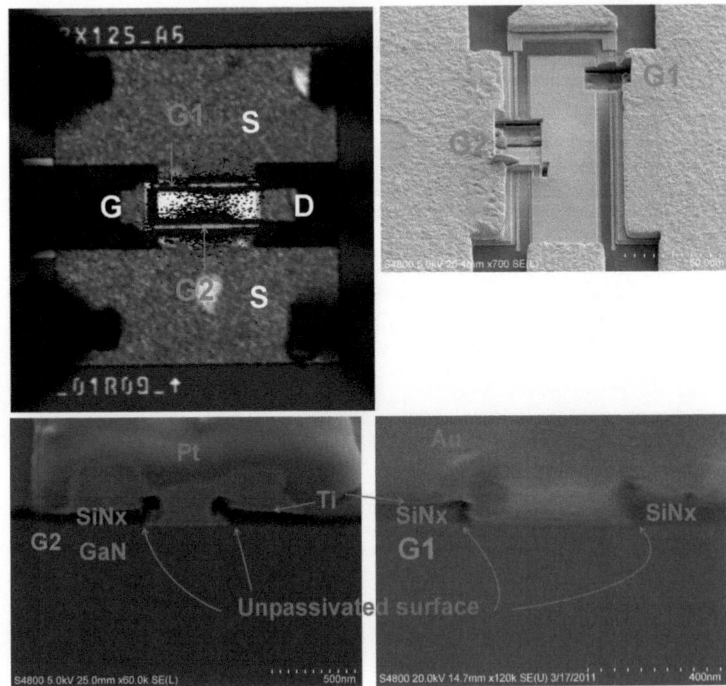

Figure 5.27 FIB cross-section of degraded device "G" at bright (G1) and dark (G2) areas at ON-state EL measurements which show voids at the gate walls and unpassivated surface [Courtesy of P. Kotara].

Chapter 6

GaN reliability interpretations

This chapter provides our general interpretations regarding the results that have been shown in chapter 5 along with interpretasions with others working in this field. It has been stated before there are two "main" streams of Al-GaN/GaN HEMTs degradation mechanisms: hot electrons (ON-state stressing) and inverse piezoelectric effect (OFF-state stressing). Both agree that the applied high electric field under the gate at the drain side region is the most critical area where the high field peak is located. This high field at ON-state stressing induces electrons in the channel to gain kinetic energy higher than lattice thermal energy, known as "hot" electrons. These hot electrons possibly can provide energy to pre-existing defects into metastable configuration that may be electrically active and migrate [38, 39]. In addition, hot electrons may release hydrogen atoms which are trapped in dislocations. Hydrogen is one of the products of GaN deposition reaction. The hydrogenation can lower the formation energy of point defects and most likely create other defects. If stressing continues, point defects may cluster to larger defective regions.

The high electric field occurance at OFF-state stress in a centrosymmetric crystal can create inverse piezoelectric effect as explained in Heckmann diagram. The AlGaN barrier layer which has a built-in tensile strain due to lattice mismatch with the GaN layer, takes up additional strain from high electric field especially during stressing in a high field range of ~ 6 MV/cm. If this total strain in AlGaN exceeds critical elasticity of the crystal, this creates crystallographic defects. This provides a pathway for electrons from the gate to penetrate in AlGaN by hopping mechanism and observed as gate leakage. However, hot electron and inverse piezoelectric effect degradation mechanism explanations do not consider the pre-existing defects i.e. point defects, dislocations and void regarding epi growth and process. Point defects such as vacancies and impurities can be the onset points for extended defects at high field stressing conditions. Moreover, point defects can be mobile and can annihilate with other vacancy, or encounter one another and therefore create more extended defect clusters. As explained in chapter 2, the formation of GaN defects is unavoidable. They can be created by the low atomic packing fraction, the stochiometry and impurity during material growth, energetic particles with lattice during etching in process steps, and different lattice and thermal expansion mismatch. It is known that point defects exist in the vicinity of dislocations. This indicates that pre-existing point defects may contribute to degradation mechanisms which is not trivial to analyse. Point defects can be observed by luminescence but dislocations are not expected, unless point defects are trapped at them due to the large stress fields near dislocations [60]. Dislocations can be measured qualitatively by XRD during epi growth or TEM after epi growth.

6.1   Nature of leakage during pinched-off stress

AlGaN/GaN HEMTs degradation mechanism due to inverse piezoelectric effect was first suggested by del Alamo group from MIT [37] who explaind the gate leakage increase during stressing through hopping mechanism (see Fig. 6.1). They observed that an abrupt increase of gate leakage current starts at a critical point where other electrical parameters such as drain resistance R_D, source resistance R_S increase abruptly and maximum drain current I_{DSS} decreases as well.

The local electric fields in a device depend on specific device design. Therefore one should be aware of using the term of critical voltage. It has to be related to the device dimension. Since we always measure the same device dimension to compare, we can use this terminology interchangebly with critical potential field. Our observations revealed that there is no meaningful difference of point of degradation between device with and without field plate although field plates are known lower the electric field at the drain side edge of the gate. This means that there must be another dominating effect causing this behavior. Interestingly, we have different result that device with GaN cap shows higher critical voltage than that of device without GaN cap [109, 112] as other group also observed [113].

During DC-Step-Stress Test at OFF-state, it has been recognized that some devices show that the gate current always increases simultaneously with the drain current (Fig. 6.2. a). This suggests the appearance of a direct leakage current path from the gate to the 2DEG region. Also surface leakage paths would explain this behavior. As soon as the device degrades in this way, the gate Schottky behaviour changes. One of the possibilities

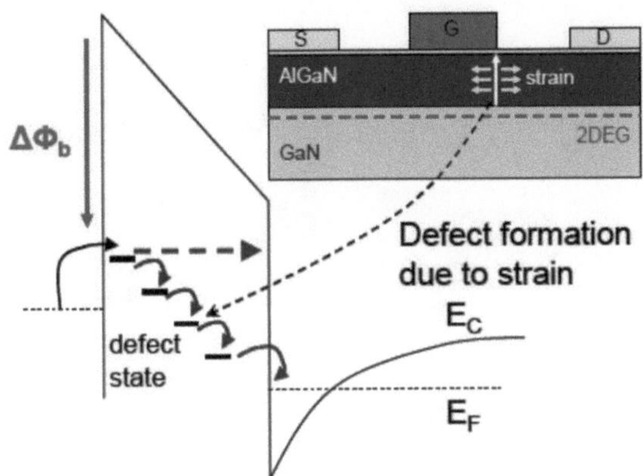

Figure 6.1 Schematic diagram of the reverse bias stress which introduces inverse piezoelectric effect. This creates additional tensile strain in AlGaN layer under the gate in the drain side. If total strain exceeds the critical values, it leads to crstallographic defects, hence electrons from the gate can penetrate in AlGaN via hopping mechanism [37].

to interpret such a behaviour is to assume a current flow in parallel to the Schottky diode. This current path can be created by initial defect generation as a consequence of high field electron injection from the gate or due to the inverse piezoelectric effect. The created defects may aggregate to larger defect clusters which then may trigger hopping conductivity from the gate region and, as finally drift to drain contact as observed as symmetric gate leakage and subthreshold drain current (see Fig. 6.2. a.).

Some devices show asymmetric behaviour of gate and subthreshold drain current after critical point during DC-Step-Stress Test at OFF-state. The subthreshold drain current I_D increases more than gate leakage I_G. We believe that this is due to punch through where electrons from the source bypass the high electric region under the gate through the buffer. This is observed as an additional subthreshold drain current I_D (see black arrows in Fig. 6.2. b.). Definitely, punch-through depends on buffer quality that one can expect it easily takes place where GaN buffer has more defects. For example device with thinner GaN buffer of 1.7 μm has dislocation density $\sim 1.4 \times 10^9$ cm^{-2}. In the high defect density regions more extended defects may be created probably due to the on-set of hopping conductivity (see black bullets in Fig. 6.2. c.) which then give rise to a large parallel current path in the degraded devices. It has been checked that parts of the devices are not bypassed by defect assisted conductivity mechanisms to the n-SiC substrate.

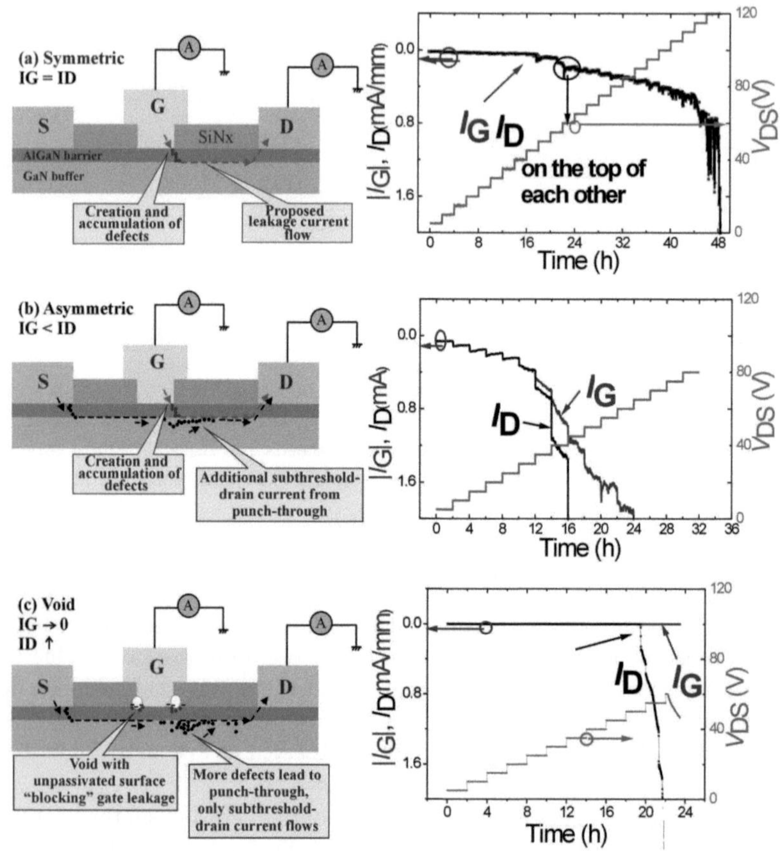

Figure 6.2 Schematic diagram charges directions of degradation points during DC-Step-Stress test at pinch-off (V_G= - 7 V) for each leakage current observation. The defects creation and/or accumulation of pre-existing defects (represented as red bullets) under the gate at the drain side due to inverse piezoelectric effect [37]. This creates a pathway between the gate and the 2DEG region. Explanation of devices with (a) symmetric, (b) asymmetric gate leakage and subthreshold drain current due to punch-through, and (c) voids at gate metal which "blocks" gate leakage current due to unpassivated surface. More defects in GaN buffer (described as black bullets) causes severe punch-through.

6.2 Electroluminescence at ON- and OFF-state

It is known that some types of dislocations are electrically active [40] and point defects can be trapped in the dislocations [60]. Those point defects can be released by high electric field during stressing. Clustered point defects may form more extended defects which may be observed as bright spots during electroluminescence measurements at OFF-state [82]. The electroluminescence of AlGaN/GaN HEMTs at ON-state originates from intraband transition was explained by Shigekawa et. al [98]. Later on they observed the correlation between high electric field and channel temperature at the edge of the gate [98].

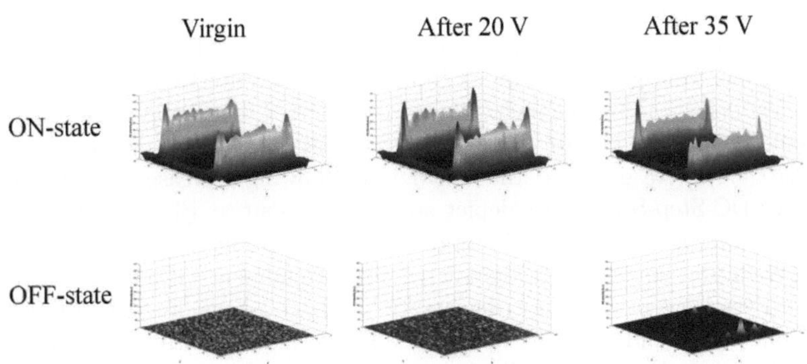

Figure 6.3 3D profiles of electroluminescence intensity along 2 finger device at ON-state (V_{GS} = -1 V) and OFF-state (V_{GS} = -7 V) with V_{DS} = 10 V)(Note:images are not yet normalized).

We observed the inhomogenousity of electroluminescence intensity along the gate fingers with 3D intensity profiles at ON-state as shown in Fig. 6.3. At the edge of both fingers show electroluminescence intensity are relatively higher because the high electric field region at these edges as compared to central device regions.

As shown before in section chapter 5, after stressing EL measurements at OFF-state show numerous bright spots broadening in size and increasesing in number where as at ON-state conditions, the EL intensity mostly decreases. Most probably, the proposed current path starts at certain regions along the gate periphery. It is therefore not homogeneous all over the gate width, at least at an early stage of degradation. Thus the increasing spotty behaviour of EL-emission during and after device degradation could be understandable as a very initial form of degradation.

6.2.1 Electroluminescence at OFF-state

Fig. 6.4 shows the schematic of heterostructure band diagram during EL measurements at OFF-state condition. For a virgin device, trapped electrons in AlGaN layer can be released by field-enhanced thermal excitation, so-called Frenkel-Poole emission (see process 2 in Fig. 6.4) [114]. EL light generation may come from native defects for example donor-acceptor pairs which are acting as potential wells for electrons. When a higher electric field is applied, these complexes which perform as dipoles, are stretched away and release electrons [82]. Electron release process can radiatively recombine to other deep level defects and emit the light and observed as bright spots at OFF-state. After stressing, more defects are created, thus more bright spots at OFF-state are observed.

During OFF-state condition (EL measurement and DC-Step-Stress Test), electrons from gate may tunnel from one defect state to another one if the defect density is high enough. This is also referred to as hopping conductivity and explains OFF-state gate leakage current I_G increase (see process 1 in Fig. 6.4). We observed that all devices showing increase of gate leakage after DC-Step-Stress-Test depict spotty behaviour at EL measurements at OFF-state condition. This supports our assumption that at early stage of degradation, some defects are created after stressing in AlGaN layer. Devices with very low gate leakage (devices "F" and "G") have practically no bright spots before and after DC-stressing. These findings correlate bright spots with gate leakage in a similar manner as observed by other group [115]. It should be noted, that devices showing a comparably low gate leakage before stressing are building up defective regions during device DC-stressing too. For example devices "G" had very low gate leakage but showed an abrupt increase of subthreshold drain current and severe performance degradation after stressing (see explanation in 6.2.2).

6.2.2 Electroluminescence at ON-state

Electrical characterization and EL-measurements suggest a degradation mechanism originating in the gate region most probably from a region near the gate edge towards the drain. Fig. 6.5 describes our understanding of EL mechanisms at ON-state before and after stressing the samples. Light generation mechanisms at ON-state originates from intra-band transitions. It has a bell-shaped EL intensity distribution over the voltage (see Fig. 5.23). As our simulation in Fig. 3.7 are showing, the electric field in 2DEG channel at ON-state is in the range of 0.4 - 0.6 MV/cm which is more than enough to excite electrons into the second conduction band of GaN (The threshold

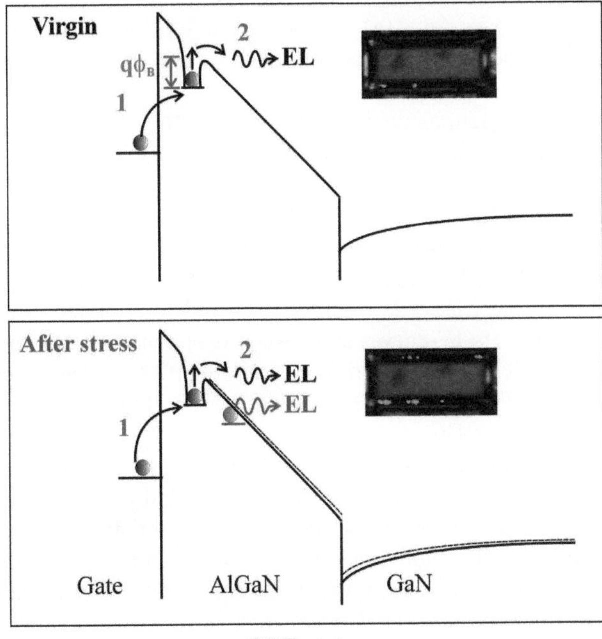

Figure 6.4 Scheme of EL mechanisms at OFF-state at virgin and after stress of a AlGaN/GaN HEMT device. Trap states have barrier height Φ_B which is the depth of the trap potential well. Bright spots already appear in virgin device as trapped electrons are released by field-enhanced thermal excitation. After stress new defects are created (represented by red circle) and contributed to wider bright spot size and new bright spots occurance. Defects creation after stress reduces the total electric field thus the conduction band slope gets lower than before. The dashed line represents the initial conduction band slope before stressing.

of electrons to occupy the second conduction band in Γ direction of wurtzite GaN bulk is 0.3 MV/cm [99]). As shown in Fig. 6.5 it is possible that electrons are excited into states high up in the second conduction band and cascade rapidly to the bottom of the second conduction band by emitting phonons (see process 1 in Fig. 6.5 (a)). Electrons are therefore able to relax to the bottom of the second conduction band before they emit photons (see process 2 in Fig. 6.5 (a)). The interplay between relaxation time after phonon emission and waiting time before emit photon emission from the bottom of the second conduction band to the first conduction band creates a thermal distribution, as depicted in Fig. 6.5 (b). This might explain heat generated in 2DEG channel of AlGaN/GaN HEMT devices during electron cascades as a scattering mechanism.

Defects created in the AlGaN layer and/or at the interface of AlGaN/GaN after stress reduce the total electric field (as shown in Fig. 6.4) and therefore the EL intensity peak drops and shifts. For moderately degraded devices, the transfer characteristics curve at $V_{DS} = 10$ V do not show any significant current drop, but its IV-curves show some drops in the knee voltage (below 10 V of V_{DS}). The Schottky diode characteristics also show some change (see Fig. 5.22). This is consistent with our assumption that defects created in AlGaN layer after DC-stressing.

For severely degraded devices, the maximum drain current drops more than 20 % due to a large number of defects created after stressing (compare the black and the red circles of transfer characteristics in Fig. 6.6). The bell-shaped curve does not dim out after stress, and interestingly, there is an additional emission at forward bias. This means that an additional EL source must have been created in the course of degradation. Most probably, defect states, energetically located in the GaN or AlGaN band gap are created. Since this effect is observed at forward bias conditons, it is reasonable that these states are located either in the AlGaN barrier or in buffer regions close to the channel (where the electron concentration is high). Such a mechanism is believed to take place in devices which may already have pre-existing defects (for example device "G" with relatively high defect density) or the injection of energetic electrons from the reverse biased gate may activate pre-existing defects and thus created extended defect clusters even in regions close to the channel.

From EL measurement results at ON-state, Zanoni *et al.* pointed out that the AlGaN/GaN HEMTs degradation mechanism is predominantly triggered by the electric field rather than by temperature [107]. They were stressing the devices at certain points along transfer characteristics. Their observation showed that if the device has been stressed at a bias point where the peak of EL occurs (still higher electric field but already rather high electron density)

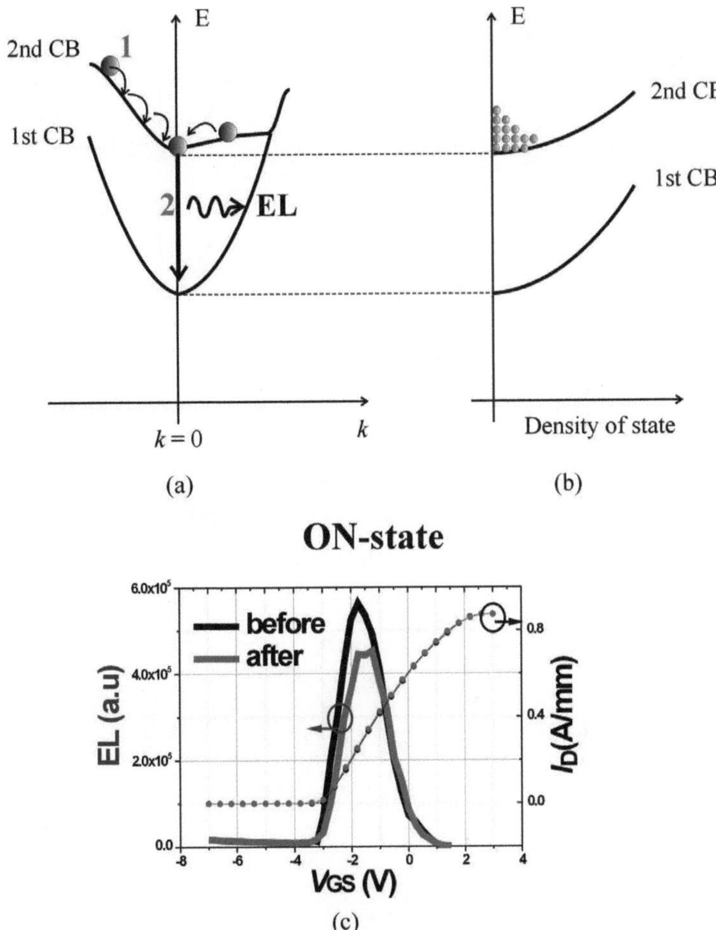

Figure 6.5 Schematic of (a) EL mechanisms at ON-state of AlGaN/GaN HEMT device due to intra-band transition. (b) Density of states and level occupancies for electrons after high electric field excitation. The red circles represent the occupancy of the available states [116]. (c) The EL intensity vs. transfer characteristics of AlGaN/GaN HEMTs at $V_{DS} = 10$ V before (black) and after (red) stress. Device has a moderate degradation. After stressing the EL intensity peak at ON-state drops and shifts due to defects creation associated herewith changes of the total electric field.

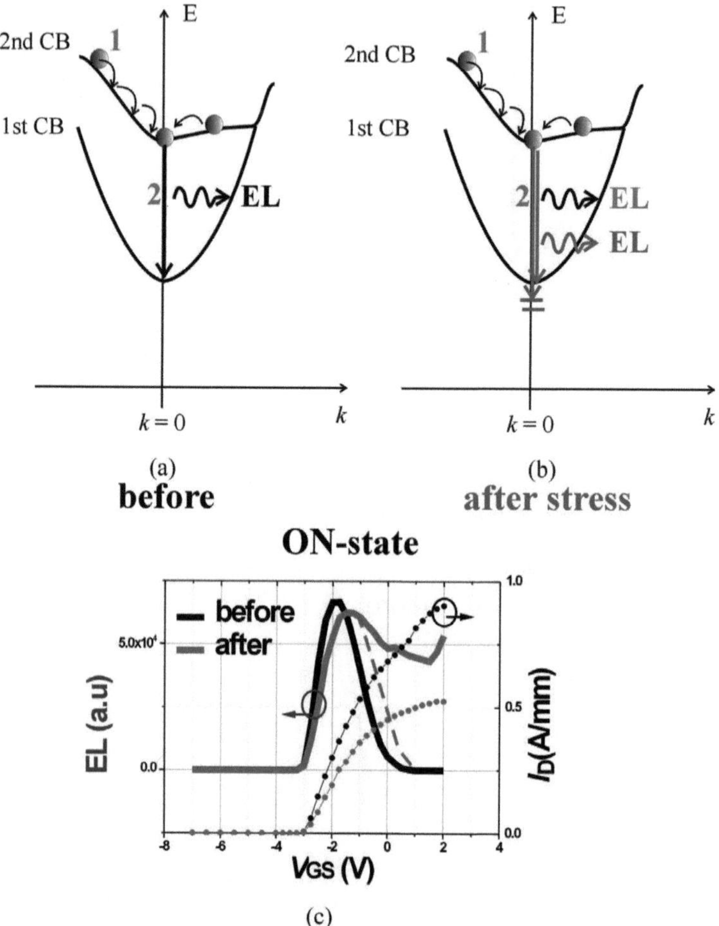

Figure 6.6 Schematic of EL mechanisms at ON-state AlGaN/GaN HEMT device (a) before and (b) after stress with additional defect states as represented by short red lines. (c) The EL intensity vs. transfer characteristics of AlGaN/GaN HEMTs with severe degradation at $V_{DS} = 10$ V before (black) and after (red) stress. An additional emission occurs at high gate voltage V_{GS} bias. The shifted bell-shaped curve is depicted by blue dashed line which originally come from blue arrow transition.

device degradation is more pronounced even as compared to stress conditions where the device has been stressed at higher power dissipation level but lower internal electric fields. It is shown that high electric field area is associated with high EL intensity [117]. This stressing at ON-state (where peak EL intensity is the maximum) convinced Ref. [107] that hot electron is the cause of degradation mechanism.

6.3   TEM investigations of dislocations

If degradation is more severe, these defects may cluster together giving an extended degraded region all along the gate edge. This region is finally characterized by interdiffusion effects as observed by Ref. [118] using TEM. Ref. [81] observed that an increased dislocation density is present under the gate area at those locations where bright spots appear during EL measurements at OFF-state conditions.

TEM supplies only very local results, i.e. a typical lamellae thickness varies from 100 to 200 nm. Furthermore, it is extremely important to indicate which imaging conditions are used to image dislocations in the layers, because only at certain imaging conditions the dislocations can be made visible. Our cross-sectioning with FIB as well as mechanical grinding and subsequent TEM investigations do not show any difference between degraded and undegraded devices. Regarding to the EL measurement results, we believe that point defect agglomerations are the best explanation for the wider size of bright spots and new bright spots occurrence after stress.

In contrast to more extended defects, point defects are not visible in standard TEM investigations. Since our investigations have been focussed to the very on-set of degradation, regions with very extensive degradation are very small in size and thus very unlikely to spot during conventional lamella preparation. All devices known from literature which show pronounced defective regions in the gate area have been stressed much more severely than our samples i.e. combination of high field and high temperature stress tests [81, 119].

Chapter 7

Conclusions and Outlook

The main focus of this thesis is to analyse AlGaN/GaN HEMTs by applying short robustness test up to the onset of device degradation. It has been found that device robustness significantly depends on epi design and material quality. Two main causes for the onset of GaN device degradation are discussed: inverse piezoelectric effect and the hot electron effect. More extended degradation effects are believed to be due to secondary effects after initial defects have been created by one of the above mentioned mechanisms.

We have developed on-wafer DC-Step-Stress-Tests at pinched-off condition by ramping drain-source voltage V_{DS} at constant rate at room temperature. During this step-stressing, high electric field is applied therefore one observes the high electric field effect on device performance solely. We defined degradation onset as the stressing condition during which either the gate leakage and/or the subthreshold drain current increases irreversible within one step. By this definition, we are able to perform fast screening tests on wafers in dependence on epitaxial designs. The device electrical characterizations i.e. IV-, DIVA, transfer and diode characteristics show different modes of irreversible degradation after DC-stressing. Further investigations to observe physical evidence were performed by SEM, FIB and TEM. All of these lead us to generate our interpretations of GaN-based device degradation mechanisms.

In order to find out which particular device region is responsible for the observed degradation effects, electroluminescence has been employed. We have shown a strong correlation between gate leakage and bright spots at OFF-state of EL measurements. Devices with very low gate leakage and AlGaN back barrier practically showed no bright spots before and after DC-stressing. This fits well to the understanding that current filament [107] most probably due to electron hopping mechanism [37] along defect structures is responsible for both, the initial gate leakage current and for the gate leakage accumulating during device degradation. These findings strongly imply how important epi quality and process technology are in determining GaN-based device reliability.

Optical charaterization by electroluminescence (EL) reveals interesting results which generate our interpretations as shown in chapter 6. Bright spots at OFF-state of EL measurements are observed and associated to point defects [82] and gate leakage [115]. After DC-Step-Stress test at OFF-state, number and size of these initial bright spots increases and broadens. Our interpretations point out that point defects agglomelerate and/or new point defects occur. During reverse bias DC-stressing, the high electric field under the gate induces the inverse piezoelectric effect or can lead to impact ionization in the high field regions which can exceed crystal elasticity and create crystallographic defects [37]. On the other hand also the injection of energetic electrons over the Schottky barrier is possible. If the electric field is high enough (if it locally exceeds the threshold for impact ionization 3.3 MV/cm), gate leakage can also be interpreted as a hole current flowing to the gate. Also these mechanisms can create localized defects. This agglomeration of these defects in AlGaN barrier creates a pathway for electrons from the gate metal to penetrate through the AlGaN barrier via hopping mechanism and observed as gate leakage. We did not observe any macroscopic defects by performing localized characterizations with SEM, FIB and TEM investigations. This supports our interpretation of point defect agglomelaration after DC-Step-Stress tests. However, Ref. [118] observed cracks under the gate after DC-stressing at 150 °C base plate temperature. Ref. [81] observed higher dislocation density where the bright spot occurs at OFF-state of EL measurements. Ref. [81] also pointed out that threading dislocation is the source of gate leakage current.

Zanoni *et al.* pointed out AlGaN/GaN HEMTs degradation mechanism is triggered by electric field rather than temperature based on EL measurement results at ON-state [107]. They observations showed devices that have been stressed at ON-state where peak EL intensity is the maximum (high electric field, low power dissipation) degraded more than that of where high power dissipation occurs (low high electric field, low EL intensity). This result convinced Zanoni *et al.* that hot electron is the cause of degradation mechanism. Stressing at OFF- and ON-states, with their well-known degradation mechanism explanations- inverse piezoelectric effect and hot electron respectively, pointed out that high electric field is the main cause of GaN device degradation.

EL peak intensity always drops and shifts to more positive gate voltage V_{GS} after DC-stressing for devices with a slight drop of drain current and/or shift threshold voltage V_{TH}. This can be a proof that our degradation criterion is good enough to detect early degradation. It is also evidence that EL is a sensitive tool to detect any small change in a device after stress.

After all, EL measurements which detect radiative recombinations (at OFF-state) and emission from intraband transition (ON-state), one should also consider non-radiative recombination occurs of minority carries in the vicinity of dislocations [120]. However one cannot avoid the pre-existing defects in a crystal which are likely the onset of degradation point.

In order to minimize defects as far as possible one must begin from the substrate i.e. pits-free and smooth surface. The most crucial part in determining epi quality is epi condition growth. In process line, one should consider that the area below the gate is the most critical point. Careful design of HEMT device is needed such as a slanted gate which not only spreads electric field but also can avoid the void at the gate walls. One should also consider radiation damage or unwanted shallow implantation of impurities which can damage semiconductor crystal and change the surface energy. In addition, SiN passivation which is intended to reduce charge surface states has to be considered especially with respect to magnitude and polarity of mechanical stress it introduces.

List of Publications/Conferences

P. Ivo, R. Pazirandeh, E. Bahat-Treidel, F. Brunner, O. Hilt, R. Lossy, J. Würfl, G. Tränkle, *GaN HEMTs critical voltage determination by DC-Step-Stress tests*, Workshop on Compound Semiconductor Devices and Integrated Circuits **51** (WOCSDICE), Leuven, Belgium, 2008 (invited).

P. Ivo, A. Glowacki, R. Pazirandeh, E. Bahat-Treidel, R. Lossy, J. Würfl, C. Boit, G. Tränkle, *Influence of GaN cap on robustness of AlGaN/GaN HEMTs*, IEEE International Reliability Physics Symposium (IRPS), p. 71-75, Montreal, Canada, 2009.

A. Glowacki, P. Laskowski, C. Boit, **P. Ivo**, E. Bahat-Treidel, R. Pazirandeh, R. Lossy, J. Würfl, G. Tränkle, *Characterization of stress degradation effects and thermal properties of AlGaN/GaN HEMTs with Photon Emission Spectral Signatures*, Microelectronics Reliability, 49, pp. 1211âĂŞ1215, 2009 (also presented in ESREF, 2009, Arcachon, France, 2009).

P. Ivo, A. Glowacki, E. Bahat-Treidel, R. Lossy, J. Würfl, C. Boit, G. Tränkle, *Comparative study of AlGaN/GaN HEMTs robustness versus buffer design variations by applying Electroluminescence and electrical measurements*, Microelectronics Reliability **51**, 217-223, 2011 (also presented in Reliability on Compound Semiconductor (ROCS) 2010, Potrland, USA, 2010).

P. Kurpas, I. Selvanathan, M. Schulz, H. Sahin, **P. Ivo**, M. Matalla, J. Splettstoesser, A. Barnes, and J. Würfl, *Stable and reproducible AlGaN/GaN-HFET processing highly tolerant for epitaxial quality variations*, Compound Semiconductor Manufacturing Technology (CS Mantech) 2010, Portland, USA, 2010.

F. Brunner, E. Bahat-Treidel, E. Cho, C. Netzel, **P. Ivo**, O. Hilt, J. Würfl, and M. Weyers, *Comparative Study of Buffer Design for High Breakdown Voltage AlGaN/GaN HFETs*, International Workshop on Nitride Semiconductors (IWN) 2010, Tampa, USA, 2010.

P. Ivo, Ute Zeimer, P. Kotara, E. Cho, L. Schellhase, E. Bahat-Treidel, J. Würfl, and G. Tränkle, A. Glowacki, and C. Boit, *Degradation mechanisms of GaN HEMTs in dependence on buffer quality and gate technology*, Reliability on Compound Semiconductor (ROCS), Palm Spring, USA, 2011.

J. Würfl, E. Bahat-Treidel, F. Brunner, E. Cho, O. Hilt, **P. Ivo**, A. Knauer, P. Kurpas, E. Lossy, M. Schulz, S. Singwald, M. Weyers, and R. Zhytnytska, *Reliability issues of GaN based high voltage power devices*, Microelectronics Reliability **51**, 1710-1716, 2011 (also presented in ESREF, 2011, Talence, France, 2011- invited).

A. Zanandrea, F. Rampazzo, A. Stocco, E. Zanoni, D. Bisi, F. Soci, A. Chini, **P. Ivo**, J. Würfl, G. Meneghesso, *DC and Pulsed Characterization of GaN-based Single- and Double-Heterostructure Devices*, Heterostructure technology (HETECH) 2011, Lille, France, 2011.

Bibliography

[1] W. C. Johnson, J. B. Parsons and M. C. Crew, "Nitrogen Compounds of Gallium III. Gallic Nitride," *The Journal of Physical Chemistry*, vol. 36, no. 10, pp. 2651–2654, 1932.

[2] H. P. Maruska and J. J. Tietjen, "The preparation and properties of vapor-deposited single-crystalline GaN," *Appl. Phys. Lett.*, vol. 15, no. 10, pp. 327–329, 1969.

[3] J.I. Pankove, H. P. Maruska, and J. E. Berkeyheiser, "Optical Absorption of GaN," *Appl. Phys. Lett.*, vol. 17, no. 5, pp. 197–199, 1970.

[4] S. Bloom, "Optical Absorption of GaN," *J. Phys. Chem. Solids*, vol. 32, pp. 2027–2032, 1971.

[5] M. Ilegems and H. C. Montgomery, "Electrical properties of n-type vapor-grown gallium nitride," *J. Phys. Chem. Solids*, vol. 34, pp. 885–898, 1973.

[6] J. Neugebauer and C. G. Van de Walle, "Atomic geometry and electronic structure of native defects in GaN," *Phys. Rev. B*, vol. 50, no. 11, pp. 867–870, 1994.

[7] C. G. Van de Walle and J. Neugebauer, "First-principles calculations for defects and impurities: Applications to III-nitrides," *J. Appl. Phys.*, vol. 95, no. 8, pp. 3851–3879, 2004.

[8] E. G. Seebauer and M. C. Kratzer, "Charged point defects in semiconductors," *Mat. Sci. Eng.*, vol. 55, pp. 57–149, 2006.

[9] I. Akasaki, H. Amano, M. Kito, and K. Hiramatsu, "Photoluminescence of Mg-doped p-type GaN and electroluminescence of GaN p-n junction LED," *J. Luminescence*, vol. 48-49, pp. 666–670, 1991.

[10] S. Yoshida, S. Misawa, and S. Gonda, "Improvements on the electrical and luminescent properties of reactive molecular beam epitaxially grown GaN films by using AlN-coated sapphire substrates," *Appl. Phys. Lett.*, vol. 42, no. 5, pp. 427–429, 1983.

[11] I. Akasaki, H. Amano,Y. Koide, K. Hiramatsu, and N. Sawaki, "Effects on AlN buffer layer on crystallographic structure and on electrical and optical properties of GaN and $Ga_{1-x}Al_xN(0{<}x{\leq}0.4)$ films grown on sapphire substrate by MOVPE," *J. Crystal Growth*, vol. 98, pp. 209–219, 1989.

[12] M. A. Khan, J. M. Van Hove, J. N. Kuznia and D. T. Olson, "High electron mobility $GaN\text{-}Al_xGa_{1-x}$ heterostructures grown by low-pressure metalorganic chemical vapor deposited," *Appl. Phys. Lett.*, vol. 58, no. 21, pp. 2408–2410, 1991.

[13] O. Ambacher, J. Smart, J. R. Shealy, N. G. Weimann, K. Chu, M. Murphy, W. J. Schaff, L. F. Eastman, R. Dimitrov, L. Wittmer, M. Stutzmann, W. Rieger and J. Hilsenbeck, "Two-dimensional electron gases induced by spontaneous and piezoelectric polarization charges in N- and Ga-face AlGaN/GaN heterostructures," *J. App. Phys.*, vol. 85, pp. 3222–3233, 1999.

[14] O. Ambacher, B. Foutz, J. Smart, J. R. Shealy, N. G. Weimann, K. Chu, M. Murphy, A. J. Sierakowski, W. J. Schaff, L. F. Eastman, R. Dimitrov, A. Mitchell, and M. Stutzmann, "Two dimensional electron gases induced by spontaneous and piezoelectric polarization in undoped and doped AlGaN/GaN heterostructures," *J. App. Phys.*, vol. 85, no. 1, pp. 334–343, 2000.

[15] R. Chu, Z. Chen, Y. Pei, S. Newman, S. P. DenBaars, and U. K. Mishra, "MOCVD-Grown AlGaN Buffer GaN HEMTs With V-Gates for Microwave Power Applications," *EEE El. Dev. Lett.*, vol. 30, no. 9, pp. 910–912, 2008.

[16] R. Chu, L. Shen, N. Fichtenbaum, D. Brown, Z. Chen, S. Keller, S. P. DenBaars, and U. K. Mishra, "V-Gate GaN HEMTs for X-Band Power Applications," *IEEE El. Dev. Lett.*, vol. 29, no. 9, pp. 974–976, 2009.

[17] T. Palacios, A. Chakraborty, S. Heikman, S. Keller, S. P. DenBaars, and U. K. Mishra, "AlGaN/GaN High Electron Mobility Transistors With InGaN Back-Barriers," *IEEE El. Dev. Lett.*, vol. 27, no. 1, pp. 13–15, 2006.

[18] Compound Semiconductor, p. 21, 2011.

[19] S. V. Novikov, C. T. Foxon, and A. J. Kent, "Growth and characterization of free-standing zinc-blende GaN layers and substrates," *Phys. Sta. Sol. A*, vol. 207, no. 6, pp. 1277–1282, 2010.

[20] J. I. Pankove, "GaN: from fundamentals to applications," *Mat. Sci. and Eng.*, no. B61-62, pp. 305–309, 1999.

[21] W. S. Yan, R. Zhang, X. Q. Xiu, Z. L. Xie, P. Han, R. L. Jiang, S. L. Gu, Y. Shi, and Y. D. Zheng, "Phenomenological model for the spontaneous polarization of GaN," *Appl. Phys. Lett.*, vol. 90, no. 182113, 2007.

[22] F. Bernardini, V. Fiorentini, and David Vanderbilt, "Spontaneous polarization and piezoelectric constants of III-V nitrides," *Phys. Rev. B*, vol. 56, no. 16, 1997.

[23] B. B. Kosicki and D. Kahng, "Preparation and Structural Properties of GaN Thin Films," *J. Vac. Sci. Tech.*, vol. 6, pp. 593–596, 1969.

[24] W. Cao, J. M. Biser, Y. Ee, X. Li, N. Tansu, H. M. Chan, and R. P. Vinci, "Dislocation structure of GaN films grown on planar and nano-patterned sapphire," *J. Appl. Phys.*, vol. 110, no. 053505, 2011.

[25] T. Sasaki and T. Matsuoka, "Substrate-polarity dependence of metal-organic vapor-phase epitaxy-grown GaN on SiC," *J. Appl. Phys.*, vol. 64, no. 9, pp. 4531–4535, 1998.

[26] R. Chierchia, "Strain and crystalline defects in epitaxial GaN layers studied by high-resolution X-ray diffraction," Ph.D. dissertation, der Universität Bremen, 2007.

[27] A. Krost and A. Dadgar, "GaN-based optoelectronics on silicon substrates," *Mat. Sci. Eng.*, no. B932, pp. 77–84, 2002.

[28] K. Hirama, Y. Taniyasu, and M. Kasu, "Dislocation structure of GaN films grown on planar and nano-patterned sapphire," *Appl. Phys. Lett.*, vol. 98, p. 162112, 2011.

[29] T. Li, M. Mastro and A. Dadgar, *III-V Compound Semiconductors-Integration with Silicon-based Microelectronics*. CRC Press, 2011.

[30] D. Ehrentraut, E. Meissner, and M. Bockoskwi, *Technology of Gallium Nitride Crystal Growth*. Springer, 2010.

[31] C. Kittel, *Introduction to Solid State Physics*. John Wiley and Sons Inc., 2005.

[32] E. F. Schubert. Band diagrams of heterostructures. [Online]. Available: http://www.ecse.rpi.edu/~schubert/ Course-ECSE-6968Quantummechanics/Ch17Heterostructures.pdf

[33] J. P. Ibbetson, P. T. Fini, K. D. Ness, S. P. DenBaars, J. S. Speck, and U. K. Mishra, "Polarization effects, surface states, and the source of electrons in AlGaN/GaN heterostructure field effect transistors," *Appl. Phys. Lett.*, vol. 77, no. 2, 2000.

[34] M. Fieger, M. Eickelkamp, L. Rahimzadeh Koshroo, Y. Dikme, A. Noculak, H. Kalisch, M. Heuken, R.H. Jansen, A. Vescan, "MOVPE processing and characterization of AlGaN/GaN HEMTs with different Al concentrations on silicon substrates," *J. Cry. Growth*, vol. 298, pp. 843–847, 2007.

[35] D. K. Sahoo, R. K. Lal, H. Kim, V. Tilak, and L. F. Eastman, "High-Field Effects in Silicon Nitride Passivated GaN MODFETs," *IEEE Trans. El. Dev.*, vol. 50, no. 5, pp. 1163–1170, 2003.

[36] G. Meneghesso,G. Verzellesi, F. Danesin, F. Rampazzo, F. Zanon, A. Tazzoli, M. Meneghini, and E. Zanoni, "Reliability of GaN High-Electron-Mobility Transistors: State of the Art and Perspectives," *IEEE Trans. Dev. Mat. Reliability*, vol. 8, no. 2, pp. 332–343, 2008.

[37] J. Joh, L. Xia, and J. A. del Alamo, "Gate Current Degradation Mechanisms of GaN High Electron Mobility Transistors," *IEDM, IEEE*, pp. 385–388, 2007.

[38] Y. S. Puzyrev, B. R. Tuttle, R. D. Schrimpf, D. M. Fleetwood, and S. T. Pantelides, "Theory of hot-carrier-induced phenomena in GaN high-electron-mobility transistors," *Appl. Phys. Lett.*, vol. 96, p. 053505, 2010.

[39] Y. S. Puzyrev, T. Roy, M. Beck, B. R. Tuttle, R. D. Schrimpf, D. M. Fleetwood, and S. T. Pantelides, "Dehydrogenation of defects and hot-electron degradation in GaN high-electron-mobility transistors," *J. Appl. Phy.*, vol. 109, p. 034501, 2011.

[40] J. E. Northrup, "Theory of intrinsic and H-passivated screw dislocations in GaN," *Phys. Rev. Lett. B*, vol. 66, p. 045204, 2002.

[41] M. G. Ancona, S. C. Binari, and D. Meyer, "Fully-coupled electrome-chanical analysis of stress-related failure in GaN HEMTs," *Phys. Stat. Sol. C*, vol. 1-2, p. 201001056, 2011.

[42] R. Quay, *Gallium Nitride Electronics*. Springer-Verlag Berlin Heidel-berg, 2008.

[43] J. F. Nye, *Physical properties of crystals*. Oxford University Press, 1957.

[44] N. Shigekawa, K. Shiojima, and T. Suemitsu, "Optical study of high-biased AlGaN/GaN high-electron-mobility transistors," *J. Appl. Phys.*, vol. 92, 2002.

[45] S. L. Chuang, *Physics of Optoelectronic Devices*. John Willey and Sons, Inc., 1995.

[46] K. Kim, D. Kim, K. Im,J. Lee, B. Kwon, H. Kwack, S. Beck, and Y. Cho, "Strong luminescence of two-dimensional electron gas in tensile-stressed AlGaN/GaN heterostructures grown on Si substrates," *Appl. Phys. Lett.*, vol. 98, p. 141917, 2011.

[47] B. Moran, F. Wu, A. E. Romanov, U. K. Mishra, S. P. Denbaars, and J. S. Speck, "Structural and morphological evolution of GaN grown by metalorganic vapor deposition on SiC substrates using an AlN initial layer," *J. Crys. Growth*, vol. 273, pp. 38–47, 2004.

[48] F. Brunner, O. Reentilä, J. Würfl, and M. Weyers, "Strain engineering of AlGaN-GaN HFETs grown on 3 inch 4H-SiC," *Phys. Stat. Sol.i C*, vol. 6, no. S2, pp. S1065–S1068, 2009.

[49] M.A. Mastro, J.R. Laroche, N.D. Bassim, C.R. Eddy Jr., "Simulation on the effect of non-uniform strain from the passivation layer on Al-GaN/GaN HEMT," *Microel. J.*, vol. 36, pp. 705–711, 2005.

[50] V. Palankovski, S. Vitanov, and R. Quay, "Field-Plate Optimization of AlGaN/GaN HEMTs," *Compound Semiconductor Integrated Circuit Symposium (CSIC), IEEE*, pp. 107–110, 2006.

[51] J. Lim, Y. Choi, Y. Kim, and M. Han, "New AlGaN/GaN HEMTs employing both a floating gate and a field plate," *Phys. Scr.*, vol. T141, no. 014009, 2010.

[52] Nidhi Chaturvedi, "Development and study of AlGaN/GaN microwave transistors for high power operation," Ph.D. dissertation, Technische Universität Berlin, 2008.

[53] L. Wang, F. M. Mohammed, and I. Adesida, "Formation mechanism of Ohmic contacts on AlGaN/GaN heterostructure: Electrical and microstructural characterizations," *J. Appl. Phys.*, vol. 103, no. 093516, 2008.

[54] S. A. Chevtchenko, P. Kurpas, N. Chaturvedi, R. Lossy, and J. Würfl, "Investigation and Reduction of Leakage Current Associated with Gate Encapsulation by SiN_X in AlGaN/GaN HFETs," *CS Mantech*, 2011.

[55] [Online]. Available: http://courses.eas.ualberta.ca/eas421/ lecturepages/microstructurediags.html

[56] E. G. Seebauer and M. C. Kratzer, *Charged Semiconductor Defects*. Springer, 2009.

[57] T. Roy, Y. S. Puzyrev, B. R. Tuttle, D. M. Fleetwood, R. D. Schrimpf, D. F. Brown, U. K. Mishra, and S. T. Pantelides, "Electrical-stress-induced degradation in AlGaN/GaN high electron mobility transistors grown under gallium-rich, nitrogen-rich, and ammonia-rich conditions," *J. Appl. Phy.*, vol. 96, p. 133503, 2010.

[58] R. T. Green, I. J. Luxmoore, P. A. Houston, F. Ranalli, T. Wang, P. J. Parbrook, M. J. Uren, D. J. Wallis, and T. Martin, "Comparison of damage introduced into GaN/AlGaN/GaN heterostructures using selective dry etch recipes," *Semiconductor Science and Technology*, vol. 24, p. 075020, 2009.

[59] J. Neugebauer and C. G. Van de Walle, *Native defects and impurities in GaN*. Festkörper Probleme: Advances in Solid State Physics Springer, 1995.

[60] M.A. Reshchikov and H. Morkoç, "Luminescence properties of defects in GaN," *J. Appl. Phys.*, vol. 97, p. 061301, 2005.

[61] J. Elsner, R. Jones, M. I. Heggie, P. K. Sitch, M. Haugk, Th. Frauenheim, S. Öberg, and P. R. Briddon, "Deep accpetors trapped at threading-edge dislocations in GaN," *Phys. Rev. B*, vol. 58, no. 19, pp. 12 571–12 574, 1998.

[62] I. Hwang, S. K. Theiss, and J. A. Golovchenko, "Mobile point defects and atomic basis for structural transformations of a crystal surface," *Science*, vol. 265, pp. 490–496, 1994.

[63] V. V. Emtsev, V. Y. Davydov, V. V. Kozlovskii, D. S. Poloskin, A. N. Smirnov, N. M. Shmidt, and A. S. Usikov, "Behavior of electrically active point defects in irradiated MOCVD n-GaN," *Phys. B*, vol. 273–274, pp. 101–104, 1995.

[64] S. Limpijumnong and C. G. Van de Walle, "Diffusivity of native point defects in GaN," *Phys. Rev. B*, vol. 269, p. 035207, 2004.

[65] M.A. Reshchikov and H. Morkoç, "Luminescence from defects in GaN," *Physica B*, no. 376–377, pp. 428–431, 2006.

[66] F. A. Ponce, D. P. Bour, and W. Götz, and P. J. Wright, "Spatial distribution of the luminescence in GaN thin films," *Appl. Phys. Lett.*, vol. 68, no. 1, 1996.

[67] S. J. Rosner, E. C. Carr, M. J. Ludowise, G. Girolami, and H. I. Erikson, "Correlation of cathodoluminescence inhomogeneity with microstructural defects in epitaxial GaN grown by metalorganic chemical-vapor deposition," *Appl. Phys. Lett.*, vol. 70, pp. 420–422, 1997.

[68] T. Hino, S. Tomiya, T. Miyajima, K. Yanashima, S. Hashimoto, and M. Ikeda, "Characterization of threading dislocations in GaN epitaxial layers," *Appl. Phys. Lett.*, vol. 76, no. 23, pp. 3421–3423, 2000.

[69] J. Elsner, R. Jones, P. K. Sitch, V. D. Porezag, M. Elstner, Th. Frauenheim, M. I. Heggie, S. Öberg, and P. R. Briddon, "Theory of Threading Edge and Screw Dislocations in GaN," *Phys. Rev. Lett.*, vol. 79, no. 19, pp. 3672–3675, 1997.

[70] H.-J. Im, Y. Ding, J. P. Pelz, B. Heying and J. S. Speck, "Characterization of Individual Threading Dislocations in GaN Using Ballistic Electron Emission Microscopy," *Phys. Rev. Lett.*, vol. 87, no. 10, p. 106802, 2001.

[71] D. C. Look and J. R. Sizelove, "Dislocation Scattering in GaN," *Phys. Rev. Lett.*, vol. 82, no. 6, pp. 1237–1240, 1999.

[72] K. Leung, A. F. Wright, and E. B. Stechel, "Charge accumulation at a threading edge dislocation in gallium nitride," *Appl. Phys. Lett.*, vol. 74, no. 17, pp. 2495–2497, 1999.

[73] D. Cherns and C. G. Jiao, "Electron Holography Studies of the Charge on Dislocations in GaN," *Phys. Rev. Lett.*, vol. 87, no. 20, p. 205504, 2001.

[74] V. M. Polyakov, V. Cimalla, V. Lebedev, K. Köhler, S. Müller, P. Waltereit, and O. Ambacher, "Impact of Al content on transport properties of two-dimensional electron gas in GaN/Al$_x$Ga$_{1-x}$N/GaN heterostructures," *Appl. Phys. Lett.*, vol. 97, p. 142112, 2010.

[75] J. W. P. Hsu, M. J. Manfra, S. N. G. Chu, C. H. Chen, L. N. Pfeiffer, and R. J. Molnar, "Effect of growth stoichiometry on the electrical activity of screw dislocations in GaN films grown by molecular-beam epitaxy," *Appl. Phys. Lett.*, vol. 78, no. 25, pp. 3980–3982, 2001.

[76] P. Kozodoy, J. P. Ibbetson, H. Marchand, P. T. Fini, S. Keller, J. S. Speck, S. P. DenBaars, and U. K. Mishra, "Electrical characterization of GaN p-n junctions with and without threading dislocations," *Appl. Phys. Lett.*, vol. 73, no. 16, pp. 975–977, 1998.

[77] E. G. Brazel, M. A. Chin, and V. Narayanamurti, "Direct observation of localized high current densities in GaN films," *Appl. Phys. Lett.*, vol. 74, no. 16, pp. 2367–2369, 1999.

[78] A. Hinoki, J. Kikawa, T. Yamada, T. Tsuchiya, S. Kamiya, M. Kurouchi, K. Kosaka, T. Araki, A. Suzuki, and Y. Nanishi, "Effects of traps formed by threading dislocations on off-state breakdown characteristics in GaN buffer layer in AlGaN/GaN heterostructure field-effect transistor," *Appl. Phys. Exp.*, vol. 1, p. 011103, 2008.

[79] H. Marchand, X. H. Wu, J. P. Ibbetson, P. T. Fini, P. Kozodoy, S. Keller, J. S. Speck, S. P. DenBaars, and U. K. Mishra, "Microstructure of GaN laterally overgrown by metalorganic chemical vapor deposition," *Appl. Phys. Lett.*, vol. 73, no. 6, pp. 747–749, 1998.

[80] Y. Xin, S. J. Pennycook,N. D. Browning, P. D. Nellist, S. Sivananthan,F. Omnés, B. Beaumont, J. P. Faurie, and P. Gibart, "Direct observation of the core structures of threading dislocations in GaN," *Appl. Phys. Lett.*, vol. 72, no. 21, pp. 2680–2682, 1998.

[81] S. Y. Park, T. Lee, and M. J. Kim, "Correlation between Physical Defects and Performance in AlGaN/GaN High Electron Mobility Transistor Devices," *Trans. El. EL. Mat.*, vol. 11, no. 2, pp. 49–53, 2010.

[82] H. Chen, Z. Lai, S. Kung, R. M. Penner, Y. Chou, R. Lai, M. Wojtowicz, and G. Li, "Observing electroluminescence from yellow luminescence-like defects in GaN high electron mobility transistors," *Jap. J. App. Phy.*, vol. 47, no. 5, pp. 3336–3339, 2008.

[83] B. Heying, E. J. Tarsa, C. R. Elsass, P. Fini, S. P. DenBaars, and J. S. Speck, "Dislocation mediated surface morphology of GaN," *J. App. Phy.*, vol. 85, no. 9, pp. 6470–6476, 1999.

[84] G. Nowak, K. Pakula, I. Grzegory, J.L. Weyher, and S. Porowski, "Dislocation Structure of Growth Hillocks in Homoepitaxial GaN," *Phys. Stat. Sol. (b)*, vol. 216, pp. 649–654, 1999.

[85] R. M. Farrell, D. A. Haeger, X. Chen, C. S. Gallinat, R. W. Davis, M. Cornish, K. Fujito, S. Keller, S. P. DenBaars, S. Nakamura, and J. S. Speck, "Origin of pyramidal hillocks on GaN thin films grown on free-standing m-plane GaN substrates," *Appl. Phys. Lett.*, vol. 96, p. 231907, 2010.

[86] T. Wernicke, S. Ploch, V. Hoffmann, A. Knauer, M. Weyers, and M.l Kneissl, "Surface morphology of homoepitaxial GaN grown on non- and semipolar GaN substrates," *Phys. Stat. Sol. (b)*, vol. 248, no. 3, pp. 547–577, 2011.

[87] H. Bang, T. Mitani,S. Nakashima, H. Sazawa, K. Hirata, M. Kosaki, H. Okumura, "Correlation between micropipes on SiC substrate and dc characteristics of AlGaN/GaN high-electron mobility transistors," *J. Appl. Phys.*, vol. 100, p. 114502, 2006.

[88] H. Sasaki, S. Kato, T. Matsuda, Y. Sato, M. Iwami and S. Yoshida, "Investigation of Surface Pits Originating in Dislocations in AlGaN/GaN Epitaxial Layer Grown on Si Substrate with Buffer Layer," *Jap. J. Appl. Phys.*, vol. 45, no. 4A, pp. 2531–2533, 2006.

[89] S. L. Selvaraj, A. Watanabe, and T. Egawa, "Influence of deep-pits on the device characteristics of metal-organic chemical vapor deposition grown AlGaN/GaN high-electron mobility transistors on silicon substrate," *Appl. Phys. Lett.*, vol. 98, p. 252105, 2011.

[90] S. L. Selvaraj, T. Suzue, and T. Egawa, "Influence of Deep Pits on the Breakdown of Metalorganic Chemical Vapor Deposition Grown AlGaN/GaN High Electron Mobility Transistors on Silicon," *Applied Physics Express*, vol. 2, p. 111005, 2009.

[91] C. P. Bayiis II, L. P. Dunleavy, "Understanding Pulsed IV Measurement Waveforms," *Electron Devices for Microwave and Optoelectronic Applications (EDMO), The 11th IEEE International Symposium*, pp. 223–228, 2003.

[92] J. I. Pankove, E. A. Miller, D. Richman and J. E. Berkeyheiser, "Electrouminescent in GaN," *J. Luminescence*, vol. 4, pp. 63–66, 1971.

[93] J. I. Pankove, *Optical processes in semiconductors.* Dover Publications, 1971.

[94] R. Gaddi, G. Meneghesso, M. Pavesi, M. Peroni, C. Canali, and E. Zanoni, "Electroluminescence Analysis of HFETs Breakdown," *IEEE Electron Device Lett.*, vol. 20, pp. 372–374, 1999.

[95] K. Hui, C. Hu, P. George, and P. K. Ko, "Impact Ionization in GaAs MESFET's," *IEEE Electron Device Lett.*, vol. 11, no. 3, pp. 113–115, 1990.

[96] N. Shigekawa, T. Enoki, T. Furuta, and H. Ito, "High-Energy and Recombination-Induced Electroluminescence of InAlAs/InGaAs HEMTs Lattice-Matched to InP Substrates," *IEEE Trans. Electron Devices*, vol. 44, pp. 513–519, 1997.

[97] G. Gradinaru, M. Asif Khan, N. C. Kao, T. S. Sudarshan, Q. Chen, and J. Yang, "Prebreakdown and breakdown effects in AlGaN/GaN heterostructures field effect transistors," *Appl. Phys. Lett.*, vol. 72, pp. 1475–1477, 1998.

[98] N. Shigekawa, K. Shiojima, and T. Suemitsu, "Electroluminescence characterization of AlGaN/GaN high-electron-mobility transistors," *Appl. Phys. Lett.*, vol. 79, pp. 1196–1198, 2001.

[99] J. Kolník, I. H. Oğuzman, K. F. Brennan, R. Wang, P. P. Ruden, and Y. Wang, "Electronic transport studies of bulk zincblende and wurtzite phases of GaN based on an ensemble Monte Carlo calculation including a full zone band structure," *J. Appl. Phys.*, vol. 78, no. 2, pp. 1033–1038, 1995.

[100] S. Yamakawa, R. Akis, N. Faralli, M. Saraniti, and S. M Goodnick, "Rigid ion model of high field transport in GaN," *J. Phys. Condens. Matter*, vol. 21, p. 174206, 2009.

[101] A. Feinberg, P. Ersland, A. Widom and V. Kaper, "Parametric degradation in transistors," *RAMS 2005*, pp. 266–270, 2005.

[102] F. Bernardini and V. Fiorentini, "First-principles calculation of the piezoelectric tensor $\overleftrightarrow{d}$ of III-V nitrides," *Appl. Phys. Lett.*, vol. 80, no. 22, pp. 4145–4147, 2002.

[103] E.T. Yu, X.Z. Dang, L.S. Yu, D. Qiaou, P.M. Asbeck, S.S. Lau, G.J. Sullivan, K.S. Boutros and J.M. Redwing, "Schottky barrier engineering in III-IV nitrides via piezoelectric effect," *Appl. Phys. Lett.*, vol. 73, no. 13, pp. 1880–1882, 1998.

[104] M. Bouya, D. Carisetti, N. Malbert, N. Labat, P. Perdu, J. C. Clément, M. Bonnet, and G. Pataut, "Study of passivation defects by electroluminescence in AlGaN/GaN HEMTs on SiC," *Microelectronics Reliability*, vol. 47, pp. 1630–1633, 2007.

[105] N. Shigekawa, T. Enoki, T. Furuta, and H. Ito, "Electroluminescence of InAlAs/InGaAs HEMTs Lattice-matched to InP Substrates," *App. Phys. Exp.*, vol. 16, pp. 515–517, 1995.

[106] E. Bahat-Treidel, O. Hilt, F. Brunner, J. Würfl, and G. Tränkle, "Punchthrough-Voltage Enhancement of AlGaN/GaN HEMTs Using AlGaN Double-Heterojunction Confinement," *IEEE Transactions on Electron Devices*, vol. 55, no. 12, pp. 3354–3359, 2008.

[107] E. Zanoni, G. Meneghesso, M. Meneghini, A. Stocco, F. Rampazzo, R. Silvestri, I. Rossetto, N. Ronchi, "Electric field and thermally activated failure mechanisms of AlGaN/GaN High Electron Mobility Transistors," *ECS 2011*, 2011.

[108] J. Joh and J. A. del Alamo, "Critical Voltage for Electrical Degradation of GaN High-Electron Mobility Transistors," *IEEE Electron Device Letters*, vol. 29, no. 4, pp. 287–289, 2008.

[109] P. Ivo, R. Pazirandeh, E. Bahat-Treidel, F. Brunner, R. Lossy, J. Würfl, G. Tränkle, "GaN HEMT critical voltage determination by DC-Step-Stress tests," *WOCSDICE*, vol. 51, 2008.

[110] R. Chu, Z. Chen, Y. Pei, S. Newman, S. P. DenBaars, and Umesh K. Mishra, "MOCVD-Grown AlGaN Buffer GaN HEMTs with V-Gates for Microwaves Power Applications," *IEEE El. Dev. Lett.*, vol. 30, no. 9, pp. 910–912, 2009.

[111] P. Ivo, A. Glowacki, E. Bahat-Treidel, R. Lossy, J. Würfl, C. Boit, G. Tränkle, "Comparative study of AlGaN/GaN HEMTs robustness versus buffer design variations by applying electroluminescence and electrical measurements," *Microelectronics Reliability*, vol. 51, pp. 217–223, 2011.

[112] P. Ivo, A. Glowacki, R. Pazirandeh, E. Bahat-Treidel, R. Lossy, J. Würfl, C. Boit, G. Tränkle, "Influence of GaN cap on robustness of AlGaN/GaN HEMTs," *IEEE International Reliability Physics Symposium (IRPS)*, pp. 71–75, 2009.

[113] P. Waltereit, S. Müller, K. Bellmann, C. Buchheim, R. Goldhahn, K. Köhler, L. Kirste, M. Baeumler, M. Dammann, W. Bronner, R. Quay, and O. Ambacher, "Impact of GaN cap thickness on optical, electrical, and device properties in AlGaN/GaN high electron mobility transistor structures," *J. Appl. Phys.*, vol. 106, p. 023535, 2009.

[114] S. M. Sze, *Physics of semiconductor devices*. John Wiley and Sons, 1981.

[115] E. Zanoni, F. Danesin, M. Meneghini, A. Cetronio, C. Lanzieri, M. Peroni, and G. Meneghesso, "Localized Damage in AlGaN/GaN HEMTs Induced by Reverse-Bias Testing," *IEEE Electron Device Letters*, vol. 30, no. 5, pp. 427–429, 2009.

[116] M. Fox, *Optical Properties of Solids*. Oxford University Press, 2001.

[117] R. J. T. Simms, "Insights into the Reliability and Physical Properties of AlGaN/GaN Based Heterostructure Devices using Optical Analysis," Ph.D. dissertation, University of Bristol, 2010.

[118] J. Joh, J. A. del Alamo, K.Langworthy, S. Xie, and T. Zheleva, "Correlation between electrical and material degradation in GaN HEMTs stressed beyond the critical voltage," *ROCS2010*, 2010.

[119] P. Makaram, J. Joh, J. A. del Alamo, T. Palacios, and C. V. Thompson, "Evolution of structural defects associated with electrical degradation in AlGaN/GaN high electron mobility transistors," *Appl. Phys. Lett.*, vol. 96, p. 233509, 2010.

[120] O. Ueda, "On Degradation Studies of III-V Compound Semiconductor Optical Devices over Three Decades: Focusing on Gradual Degradation," *Jap. J. Appl. Phys.*, vol. 49, p. 090001, 2010.

Innovationen mit Mikrowellen und Licht
Forschungsberichte aus dem Ferdinand-Braun-Institut
Leibniz-Institut für Höchstfrequenztechnik

Herausgeber: Prof. Dr. G. Tränkle, Prof. Dr.-Ing. W. Heinrich

Band 1: **Thorsten Tischler**
Die Perfectly-Matched-Layer-Randbedingung in der
Finite-Differenzen-Methode im Frequenzbereich:
Implementierung und Einsatzbereiche
ISBN: 3-86537-113-2, 19,00 EUR, 144 Seiten

Band 2: **Friedrich Lenk**
Monolithische GaAs FET- und HBT-Oszillatoren
mit verbesserter Transistormodellierung
ISBN: 3-86537-107-8, 19,00 EUR, 140 Seiten

Band 3: **R. Doerner, M. Rudolph (eds.)**
Selected Topics on Microwave Measurements,
Noise in Devices and Circuits, and Transistor Modeling
ISBN: 3-86537-328-3, 19,00 EUR, 130 Seiten

Band 4: **Matthias Schott**
Methoden zur Phasenrauschverbesserung von
monolithischen Millimeterwellen-Oszillatoren
ISBN: 978-3-86727-774-0, 19,00 EUR, 134 Seiten

Band 5: **Katrin Paschke**
Hochleistungsdiodenlaser hoher spektraler Strahldichte
mit geneigtem Bragg-Gitter als Modenfilter (α-DFB-Laser)
ISBN: 978-3-86727-775-7, 19,00 EUR, 128 Seiten

Band 6: **Andre Maaßdorf**
Entwicklung von GaAs-basierten Heterostruktur-Bipolartransistoren
(HBTs) für Mikrowellenleistungszellen
ISBN: 978-3-86727-743-3, 23,00 EUR, 154 Seiten

Band 7: **Prodyut Kumar Talukder**
Finite-Difference-Frequency-Domain Simulation of Electrically
Large Microwave Structures using PML and Internal Ports
ISBN: 978-3-86955-067-1, 19,00 EUR, 138 Seiten

Band 8: **Ibrahim Khalil**
Intermodulation Distortion in GaN HEMT
ISBN: 978-3-86955-188-3, 23,00 EUR, 158 Seiten

Band 9: **Martin Maiwald**
Halbleiterlaser basierte Mikrosystemlichtquellen für die Raman-Spektroskopie
ISBN: 978-3-86955-184-5, 19,00 EUR, 134 Seiten

Band 10: **Jens Flucke**
Mikrowellen-Schaltverstärker in GaN- und GaAs-Technologie
Designgrundlagen und Komponenten
ISBN: 978-3-86955-304-7, 21,00 EUR, 122 Seiten

Cuvillier Verlag
Internationaler wissenschaftlicher Fachverlag

Innovationen mit Mikrowellen und Licht
Forschungsberichte aus dem Ferdinand-Braun-Institut
Leibniz-Institut für Höchstfrequenztechnik

Herausgeber: Prof. Dr. G. Tränkle, Prof. Dr.-Ing. W. Heinrich

Band 11: **Harald Klockenhoff**
Optimiertes Design von Mikrowellen-Leistungstransistoren
und Verstärkern im X-Band
ISBN: 978-3-86955-391-7, 26,75 EUR, 130 Seiten

Band 12: **Reza Pazirandeh**
Monolithische GaAs FET- und HBT-Oszillatoren
mit verbesserter Transistormodellierung
ISBN: 978-3-86955-107-8, 19,00 EUR, 140 Seiten

Band 13: **Tomas Krämer**
High-Speed InP Heterojunction Bipolar Transistors
and Integrated Circuits in Transferred Substrate Technology
ISBN: 978-3-86955-393-1, 21,70 EUR, 140 Seiten

Band 14: **Phuong Thanh Nguyen**
Investigation of spectral characteristics of solitary
diode lasers with integrated grating resonator
ISBN: 978-3-86955-651-2, 24,00 EUR, 156 Seiten

Band 15: **Sina Riecke**
Flexible Generation of Picosecond Laser Pulses in the Infrared and
Green Spectral Range by Gain-Switching of Semiconductor Lasers
ISBN: 978-3-86955-652-9, 22,60 EUR, 136 Seiten

Band 16: **Christian Hennig**
Hydrid-Gasphasenepitaxie von versetzungsarmen und freistehenden
GaN-Schichten
ISBN: 978-3-86955-822-6, 27,00 EUR, 162 Seiten

Band 17: **Tim Wernicke**
Wachstum von nicht- und semipolaren InAlGaN-Heterostrukturen
für hocheffiziente Licht-Emitter
ISBN: 978-3-86955-881-3, 23,40 EUR, 138 Seiten

Band 18: **Andreas Wentzel**
Klasse-S Mikrowellen-Leistungsverstärker mit GaN-Transistoren
ISBN: 978-3-86955-897-4, 29,65 EUR, 172 Seiten

Band 19: **Veit Hoffmann**
MOVPE growth and characterization of (In,Ga)N quantum structures
for laser diodes emitting at 440 nm
ISBN: 978-3-86955-989-6, 18,00 EUR, 118 Seiten

Band 20: **Ahmad Ibrahim Bawamia**
Improvement of the beam quality of high-power broad area
semiconductor diode lasers by means of an external resonator
ISBN: 978-3-95404-065-0, 21,00 EUR, 126 Seiten

Cuvillier Verlag
Internationaler wissenschaftlicher Fachverlag

Innovationen mit Mikrowellen und Licht
Forschungsberichte aus dem Ferdinand-Braun-Institut
Leibniz-Institut für Höchstfrequenztechnik

Herausgeber: Prof. Dr. G. Tränkle, Prof. Dr.-Ing. W. Heinrich

Band 21:   **Agnietzka Pietrzak**
Realization of High Power Diode Lasers with Extremely Narrow
Vertical Divergence
ISBN: 978-3-95404-066-7, 27,40 EUR, 144 Seiten

Band 22:   **Eldad Bahat-Treidel**
GaN-based HEMTs for High Voltage Operation
Design, Technology and Characterization
ISBN: 978-3-95404-094-0, 41,10 EUR, 220 Seiten

Cuvillier Verlag
Internationaler wissenschaftlicher Fachverlag